PATTERNS

in the

SAND

FRONTIERS OF SCIENCE
Series editor: Paul Davies

COMPUTERS, COMPLEXITY, AND EVERYDAY LIFE

PATTERNS
in the
SAND

TERRY BOSSOMAIER • DAVID GREEN
FOREWORD BY PAUL DAVIES

§

HELIX BOOKS

PERSEUS BOOKS
Reading, Massachusetts

Library of Congress Catalog Card Number: 98–86413

ISBN 0-7382-0015-8

Published in Australia by Allen & Unwin Pty Ltd

Perseus Books is a member of the Perseus Books Group

Jacket design by Suzanne Heiser
Set in 10.5-point Plantin by DOCUPRO, Sydney

1 2 3 4 5 6 7 8 9–DOH–0201009998
First printing, July 1998

Find Helix Books on the World Wide Web at
http://www.aw.com/gb/

CONTENTS

FOREWORD

A casual glance at the world around us reveals such a rich diversity of physical forms and systems it seems incredible that we could ever come to understand nature. The Greek philosophers struggled with this challenge and one of their number, Democritus, proposed a neat solution. The entire universe, he maintained, was composed of nothing but atoms moving in a void. In this way, the huge variety of physical things might be attributed to the unlimited number of ways in which these atoms could combine and, therefore, everything that happens in nature would be due to the rearrangement of these atoms. The immense complexity of the physical universe could then be reduced to the antics of a few species of basic building blocks.

The reductionist world view that stemmed from Democritus' atomism was destined to become a paradigm for science. Nature may be bewilderingly complex, it was said, but if only the nature of the building blocks could be understood, then all would be explained, for the whole would be nothing more than the sum of its parts. It is an enticing philosophy, however, reductionism has been roundly condemned for committing the fallacy of 'nothing-buttery'. Take a human being. Since people are made up of cells, the reductionist seeks to explain human behaviour by understanding how cells (including brain cells) work. But cells are just complicated bags of molecules, so their properties should be explicable by chemistry. Chemistry, in turn, is just a branch of physics, so we should, in principle, be able to derive the chemical properties of bio-molecules from the laws of atomic and subatomic physics. Thus psychology and biology reduce to chemistry, which in turn reduces to particle physics. If this chain of reasoning were correct, it would mean that, ultimately, all human qualities could be embedded in the properties of the underlying building blocks—the fundamental particles and fields out of which the world is put together.

While reductionism remains a powerful methodology, few scientists believe it is the whole story. Over the last century, physics has cured itself of its once strong reductionist flavour, as more and more phenomena have been discovered that don't fit comfortably into the paradigm. Historically, one can see glimmerings of trouble for reductionism in the work of Henry Poincaré, a French mathematical physicist who, at the turn of the century, discovered what we would now call the theory of chaos. Before this, it was generally supposed that complexity is mere complication caused by the sheer number of participating units. No fundamentally new laws or principles were expected to be involved. Thus, a biological organism or a snowflake or a turbulent stream may appear complex, but this is only because we do not know the details of all the tiny components that constitute them nor the way that they interact. It was generally supposed that beneath the surface complication of such systems, there must lie simplicity at the level of individual components. Chaos theory demolished this belief by showing that even simple systems with very few components can be so exquisitely sensitised to the tiniest of disturbances that they behave in an essentially random and unpredictable manner.

vii

But this was only the beginning. Other forms of complexity, where the random anarchy of chaos is replaced by more constructive activity, were discovered. The laser and the superconductor showed that countless independent elements can act collectively in a cooperative manner, as if guided by an unseen hand. Even a humble pan of hot water can spontaneously arrange itself into a pattern of hexagonal convection cells. The terms 'synergetics' and 'self-organisation' were coined to describe the coherent activity of complex systems that display long-range cooperative behaviour. Sometimes assemblages of simple, unexciting components will take on a life of their own and act in surprising and even creative ways that are impossible to predict. Often small external changes will trigger major consequences as a system reaches an instability that may lead either to a catastrophe or mark the transition to a state of still more elaborate complexity.

But how can scientists understand these complex systems? The problem about complexity is that, by its very nature, it seems impossible to grasp in generality. No two cloud patterns, planetary rings or biological organisms are the same. However, with the advent of the electronic computer, researchers were able to study a vast range of complex behaviour. They started to notice similarities in the structure and activity of widely different complex systems. Universal mathematical patterns were discerned even in chaotic systems. It began to look as if there might be quite general laws of complexity to place alongside the laws of physics.

A completely new culture began to emerge among scientists studying such complex systems. The reductionist paradigm is founded on the idea of treating the smallest components as primitive, irreducible and simple, then trying to explain complex systems by the combined activity of their myriad parts. Complexity theorists turned this on its head. Instead of a bottom-up approach, they sought examples of primitive, irreducibly complex components. One such is the fractal, a geometrical object that is in some sense infinitely irregular (just as Democritus' atoms were infinitely simple). Nevertheless, fractals have a very elegant and manageable mathematical description and so can be used to model complexity very efficiently. An often-cited example is the coastline of Great Britain, which is wiggly on all scales of size, but in a way that has a simple underlying scaling relation. Indeed, it is often cheaper on data to fractalise the image of a natural scene rather than to pixelate it (which is the reductionist's approach). This is because nature is full of fractal patterns—not only coastlines, but also mountain ranges, forests and river valleys, to name but a few.

Another important concept is the network—a collection of objects connected in a way that they can influence each other to a greater or lesser extent. Networks are familiar from the structure of brains, the evolution of ecosystems and the movements of markets. They occur in magnetic materials and immune systems, in ant colonies and gene interactions and, most famously, in the Internet and World Wide Web. By definition, a network cannot be reduced to its components, for it is its very inter-connectedness that bestows upon a network its distinctive qualities. Once again, researchers see a hint of universal principles at work. They glimpse self-organisation and chaos in systems as diverse as currency markets and rainforests and the patterns that decorate the surfaces of sand piles. Their work holds out the tantalising promise that aspects of nature (or even human behaviour) which once appeared incomprehensible and utterly intractable might yet be accurately modelled using new mathematical principles.

Terry Bossomaier and David Green belong to this new breed of researcher. Their expertise transcends traditional subject boundaries. Together, they bring to the study of complexity a background of biology, computing, neural nets and mathematics. While sceptical of some of the wilder claims made for complexity theory, they are convinced that we stand on the threshold of a new science that promises to have wide-ranging implications for our understanding of nature. In this book, they describe many everyday examples of the complexity theorist's craft at work. Their account leaves no doubt that the subject is not just one of grandiose visions and paradigm-busting leaps: it is also about down-to-earth practical applications. Take, for example, their discussion of accidents and criminal responsibility, or the spread of epidemics—both cases where increasing social complexity can cause escalating problems and instabilities.

When I was a student, it was often said that the two great frontiers of science were the very large and the very small—cosmology and particle physics. Today there is a third frontier, equally challenging and important, that of the very complex. There are many occasions in the history of science where the word 'revolution' has been glibly applied. True scientific revolutions are actually rather rare, however, and involve more than sudden advances in this or that discipline. The complexity revolution is aptly named, for it goes beyond merely providing interesting examples: it changes the way that scientists approach their subject matter. It has given us new tools and concepts, as well as fresh explanations for age-old puzzles. Mankind has always known that the physical universe is complex—magnificently, exuberantly, richly complex—but only now are we beginning to understand just how subtle and extensive that complexity may be.

Paul Davies
Adelaide, 1997

FIGURES

GLOSSARY

algorithm A recipe for performing a function or solving a problem.

alife (artificial life) The study of life-like properties in artificial systems.

analog/digital Analog processes use 'analogies' of physical phenomena (e.g., electrical voltage for the size of a variable), whereas digital procedures represent phenomena through numbers or symbols. Most computers today are digital, but analog computers are still used.

attractor A pattern that 'attracts' a process. Fractals are the attractors of particular drawing algorithms. Equilibria are the point attractors for physical processes.

carbon cycle The sequence of steps that carbon cycles through an ecosystem. For example, plants fix carbon in their tissues, herbivores eat the plants, carnivores eat the herbivores, bacteria decompose the dead bodies of carnivores when they die, thereby passing the carbon into the soil so that plants can extract it to begin the cycle again.

cellular automaton (CA) An array in which each cell is an automaton (computer) with identical programming. Each cell interacts with neighbouring cells. The most famous example is John Conway's game of LIFE.

chaos Complex behaviour in dynamics systems. One characteristic is sensitivity to initial conditions. That is, small initial differences quickly lead to large differences in behaviour. Chaotic systems travel through many states without ever returning to their starting point.

codon A string of three DNA bases that code for a single amino acid.

computability The existence of an algorithm to solve a problem. See also **halting problem**.

cyberspace Thought space, the universe of information and ideas. Derives from Norbert Wiener's term 'cybernetics'. The concept of cyberspace can be compared to the philosopher Karl Popper's idea of a third world.

DNA/RNA (Deoxyribose or ribonucleic acid) The chemicals that store the genetic code.

DLA (diffusion limited aggregation) Formation of patterns or structures by aggregation in which the particles have to diffuse through a medium before sticking to a developing cluster.

disorder The term is used in a fairly precise way by physicists and engineers to mean a system with a lot of inherent randomness and therefore low predictability.

dissipative system A system that exchanges energy with its surroundings.

domain A logical section or grouping of sites on the

Internet, such as a country (e.g., '.uk' indicates Great Britain) or a type of organisation (e.g., '.com' for commercial sites).

emergence The appearance of large-scale properties in self-organising systems; e.g., seeing the wood for the trees.

entropy A measure of disorder in a system. A fundamental concept in thermodynamics and information theory.

environmental gradient A systematic change (across space or time) of an environmental property (e.g., soil moisture or temperature). An example is the drop in temperature noticeable as you climb a mountain.

fractal A pattern or process made up of many replicas at different scales. If all are the same it is said to be self-similarity, but if they are the same only on average, it is said to be self-affined.

graph Mathematicians have a precise meaning for a graph, somewhat different from everyday language. A graph is a set of points (nodes) some or all of which are connected by lines (edges).

halting problem The question of deciding whether a given Turing Machine program will ever finish. In general, the problem cannot be solved, that is, we cannot always tell whether or not a problem can be solved. A Turing Machine that will stop defines an algorithm for solving a problem.

hypercycle A cyclic pattern of chemical reactions in which each substance catalyses the formation of later substances. Hypercycles are thought to have played a part in the origin of life.

IP address The location of a computer. This is used by the Internet Protocol (IP) to pass messages to that computer.

iteration Repetition of a process again and again. Fractals are formed by iteration of the same process at ever smaller scales.

loop analysis This is a method of analysing the feedback loops in a complex dynamic system by tracing the signs of the interactions.

mapping A transformation that links input values to output values. For example, adding 1 to any number defines a mapping from each number X to $X + 1$; i.e., it maps the number 1 to 2, 2 to 3, and so on.

non-linearity The presence of non-linear terms or components (e.g., square, cubic) in a system or process. Non-linearity is essential for chaos. Interactions in physical systems are often non-linear.

phase-locking Constraining different processes to cycle in unison; e.g., clocks on a wall interact so that they all keep identical time.

phase space An imaginary space in which each dimension represents a variable or parameter of the system.

pixel A picture element. We can generate pictures as large chequerboards in which each square (pixel) is a different colour. Images on a computer screen are composed of thousands of pixels.

protocol Conventions for computers to exchange information over a network.

random variations The term derives from communications in which an orderly signal can be obscured by random background noise.

reductionist paradigm An approach to science in which we try to understand complicated phenomena by breaking them down into simpler parts. For example, to try to understand the human body we look at the functions of different organs and muscles.

stochastic Random values. For example, a stochastic process takes random values at any time.

synergetics The name Haken gave to phenomena in which lots of objects coordinate their behaviour and act in unison (e.g., a laser firing).

Turing Machine A hypothetical ultra-simple computer proposed by Alan Turing as a model for computability.

turtle geometry A method of drawing pictures by following the graph of an imaginary turtle around a plane invented by Seymour Papert.

UGC (unique giant component) A cluster of connected nodes in a graph that absorbs most of the nodes.

universal computation The ability of a computing system to carry out any computing task. It is often used in the sense of a Universal Turing, a Turing Machine that can simulate all other Turing Machines.

PREFACE

This book distils many years of practical experience with complexity. Our work on neural systems, environmental processes and many other questions in biology and computing has taught us the importance of interactions and other mechanisms that produce complexity. We want to share with you, the reader, some of the excitement we feel about this profound and fascinating new field of knowledge.

As scientists who apply computers to biology all the time, we have seen over the years the growing importance of the computer, both as a practical, working tool and as a metaphor for many aspects of the living world. Strange as it may seem, biology and computing are today very closely related disciplines. Not only is computing teaching us a lot about biology, computer scientists are also learning a lot from seeing how nature has dealt with many issues in computation and information processing. We try to draw out this close relationship again and again in different contexts throughout the book.

Complexity is an extremely rich area of research that draws together ideas, methods and phenomena from almost every traditional field of knowledge. Unlike recent popular books in this area, we do not dwell on the personalities of the scientists doing the research. Nor do we try to be encyclopaedic. Of necessity, there are many topics that we deal with only in passing, or not at all.

Instead, the main focus of the book is on what complexity is and how it impinges on our everyday lives. We use a lot of concrete examples—from starfish to travelling salesmen, from car crashes to the workings of the brain. Our hope is that our readers will go away with an appreciation of the main issues. If we help you to see the world in a new light, then we have succeeded. Finally, we invite readers who want to explore the world of complexity further to browse the on-line resources and information available through Complexity On-Line. This service can be located via our Web server at the Internet address http://life.csu.edu.au.

Terry Bossomaier
David Green
December 1997

CHAPTER 1

COPING WITH COMPLEXITY

The date was 5 May 1961. At Cape Canaveral, Florida, the United States' first manned space flight was about to get under way. Out on the launch pad, astronaut Alan B. Shepard was strapped into his Mercury capsule on top of a Redstone missile. This first flight was to be quick—an up-and-down, sub-orbital trip just beyond the edge of space.

Things were not going well. Time and time again engineers halted the countdown to check possible faults in one system after another. The problem was that everyone was dead scared of failure. After the Russians successfully sent Yuri Gagarin into space in April there had been tremendous pressure to get an American up there too. But American rockets had a history of blowing up in public. A similar disaster on this first manned flight would have seen American prestige and credibility blow up with the rocket.

As a result, every technician was frightened of fouling up. No one wanted to be the one who caused a disaster. So they checked and rechecked every minute detail of the systems under their control. The rocket could not be launched until every system had been confirmed as 'go'. But no one wanted to commit themselves in case they had missed some fault.

Although the flight was to last only fifteen minutes,

1

Shepard had been sealed in his capsule for over four hours. By this time his bodily functions began to take priority. The coffee he drank for breakfast had passed through his digestive system and wanted desperately to escape. No one knew what the effect would be of a sudden release of bodily fluids into his spacesuit. Shepard's body was taped with biosensors. Fluid could spark an electrical shortcircuit. A major concern was fire. Six years later three astronauts would be killed when a fire raged inside the pure oxygen environment of the first Apollo capsule.

Eventually he could hold on no longer. Instruments went crazy and a lot of biomedical data were lost. The go-ahead to launch was given before there were any more problems. At long last America's first manned space flight got under way.[1]

The point of this story is that a spaceship is an excellent example of a complex system. Thousands of separate systems have to work correctly if the spacecraft is to fly. With careful attention to every minute detail the risk of any one part failing is very small. However, there are literally millions of individual parts. So the overall odds of some part failing is depressingly high.

Another familiar example of complex systems is home electronics. Today electronic devices tend to be very reliable. Many TV sets, for instance, run trouble-free for years. This reliability is a result of advances in electronics. In the early days, TVs consisted of dozens of vacuum tubes. Like light bulbs, these valves tended to fail after a time, making visits by the TV repairman all too familiar. Unreliable vacuum tubes caused far more severe problems for the first computers. Whereas a TV set had perhaps a few dozen valves, early computers contained thousands. It took a large room full of valves and wires to build a device that was hardly more powerful than a pocket calculator. Failures were so frequent that valves were constantly being replaced. It was a challenge to make the machines work at all.

The story of computer programming is similar. Computer programs can be extremely complex. A major piece of software can contain many thousands of lines of code. In the early days of computing, the programmer would sit down and carefully plan out a program and draft the code on paper. This written version would then be typed up as a series of punched cards—one card per line of code. For a typical program these cards would fill a box. To run the program you would deposit the box at the computer centre and come back the next day to pick up the output. Inevitably the output would contain errors, so you would spend the afternoon poring over the code to debug the source and replace some of the cards before resubmitting the job yet again.

One rule of thumb in programming is that the number of bugs in a piece of new code increases exponentially with the number of lines. The shorter the segment of code the better your chances of getting it right. At first that doesn't make sense: the number of bugs in a few lines of code should be the same regardless of the size of the rest of the program. But what happens is that each new line can interact with existing lines of code in obscure and usually troublesome ways. Today this rule leads to the practice in which programmers write small additions to their code, then compile, test and debug all within minutes. In the early days, however, that approach meant months of painstaking development. In a later section we discuss a new trend in computing—a trend not only growing very fast but returning huge gains in productivity. The idea is crucial to managing complexity: encapsulating data with the procedures that can access and manipulate it.

Just as engineers developed procedures to reduce the risk of errors, so programmers have developed techniques

3

to minimise coding errors. Whereas engineers divide a complex machine into subsystems, programmers organise their programs into modules. Each module can be developed and tested before moving on to the next one. This approach really becomes powerful when you start reusing modules again and again, as building blocks. Its potential was first realised early in the nineteenth century by Ada Lovelace, the world's first programmer, who developed methods for Babbage's Analytic Engine.[2]

When the first modern computers were built after World War II, programs were all written in what is known as machine code. These were instructions that told the computer literally what to do. Instructions at this level are extremely detailed and convoluted. They are full of obscure commands like 'move the number in register 1 to register 2'. Perhaps the greatest advance in computing was to develop ways of giving names to groups of commands that are used again and again. This practice led to a special kind of program, called a compiler, that converted all those names of routines into the appropriate machine code. In this way programming languages were born.

Engineers adopt many strategies to avoid catastrophic failures. The most visible are monitors that report on operation of the system. We are all familiar with speedometers and fuel gauges in our cars. Complex machines such as airliners and spaceships need to have many times more sensors than a car has. The advantage of sensors is that problems can often be identified before they become catastrophic. However, each sensor is itself a new system that must also operate correctly. One of the most common problems in spaceflight is faulty indicator lights.

Another strategy is to build tolerances into parts and systems so that they can operate well beyond the range normally expected. A dramatic example of the dangers of poor tolerances was the Challenger disaster, in which a space shuttle blew up shortly after liftoff, killing the seven astronauts on board. The booster's O-ring seals failed because they became brittle when used in temperatures below those they were designed to tolerate.

Redundancy is yet another approach to coping with part failure. If a system is crucial you duplicate it. The space shuttle, for example, has three computers on board.

Without all of the above precautions, and many more, challenging activities, like flight or space travel, would be impossible. And yet, even after all those precautions, perhaps the biggest problem of all remains. And it is that the interactions of different systems with each other are the greatest source of complexity. Just look under the bonnet of your car. The packing of all the complex parts under the hood is a triumph of design. Change anything and that change affects everything else. Enlarge the carburettor and the exhaust hose no longer fits. Move the exhaust hose and it could overheat the electrical wiring.

Yet, despite all these techniques and precautions, many of us have had the experience of staring in frustration at the computer screen after something has gone wrong. Maybe the machine has just frozen, maybe a file you've been working on for the last hour has disappeared—whatever it is, it's usually annoying! We don't have a clue what has happened, and we don't need to be told that computers are complex. Sometimes the behaviour is really spooky. Installing a new software package causes tried and trusted programs to play up. Yet today's digital computer is quite a simple rule-based device, with nothing like the sophistication of a biological brain. These screen lock-ups or other frustrations arise usually through failures in a program or sometimes in the interaction between programs, perhaps in the same computer, perhaps over the Internet.

The possible complications of interactions between systems are endless, and often unpredictable. That is why a completely unforeseen situation, such as Shepard's fall from grace, is always on the cards in spaceflight. It is also the biggest headache for designers of any large system. The packing of a car's engine under the bonnet did not just happen overnight. It is the end product of literally decades of gradual improvements and design changes.

FROM CATS TO THE WEB OF COMPLEXITY

It's as difficult as herding cats, was how a frustrated dean once described running a university faculty. Cats are independent creatures and hard to control. Getting lots of them to be well behaved is impossible. But you don't have to be as complex or independent an entity as a cat to get an unpredictable system. Isaac Newton used to complain that thinking about the moon made his head ache. The reason? Try as he might he just couldn't solve the equations of motion for the system of the earth, moon and sun. What he did not know (it was not proved until several centuries later) was that his efforts were doomed. No solution exists for the so-called *three body problem*, of which this is a special case.

Nor do you need many cats for chaos, either, as a kitten with a ball of wool will soon demonstrate! As the kitten chases the ball around the house it unravels, leaving chairs, tables, just about everything, tied together. Each time the ball hits an obstacle it bounces off at an odd angle, since it is not a perfect sphere. Unless we can find an end to the wool, we have no idea where the path started or stopped. Here we have one agent, following a complicated path, going round and round the same area along slightly different paths each time.

We might call this the *iterative* approach to complexity, the repetition of simple acts again and again.

A spider's web is another example of a complex pattern formed by a single agent. The spider traces a complicated path in time as it spins: the pattern of the resulting web in space is a record of the spider's path. We can see here a very important idea which we will meet frequently. Complexity is seen *not at the level of an individual*, be it animal, plant or mineral, *but in the larger picture*, which may be an aggregate view over time or space. In the case of the spider's web we can see a very clear pattern; in the case of the ball of wool there is no obvious pattern at all. But patterns, intertwined paths, structures of all kinds, are hallmarks of complexity.

So just what is complexity?

In essence, complexity is about the ways in which the world is put together. As the above examples already show, when you put lots of things together, such as subsystems in an airliner, the interactions between them can be extremely complex indeed. It is these interactions that turn something from being merely complicated (having many elements) into something truly complex.

If you roll a billiard ball across a billiard table, then any good billiards player can predict the path that it will follow. If you roll two balls, then the problem is still simply a matter of calculating each ball's path individually. However, there is also a chance that the two balls will bump into each other. Add a few more balls and the problem of keeping track of paths and collisions becomes very difficult and predicting the state of all the balls becomes well nigh impossible.

Suppose now that you roll (say) 100 balls around on the table. With so many balls, they will be bumping into each other all the time. The problem of computing

7

individual paths becomes virtually impossible. Such a system is essentially unpredictable at the individual level. It is hard to record, in the sense of logging where everything is, and devoid of simple patterns. However, if what we want to know is not the path of any particular billiard ball but the appearance of the entire table, then with large numbers of balls the problem starts to become simpler again. We no longer need to trace every ball exactly. Instead we can look at average behaviour. The myriad individual interactions average out and we can make sensible predictions about average speed, average time between collisions and the average distribution of balls on the table.

Here we have an example of a key idea in complexity: *emergence*. The average properties described above *emerge* out of countless individual interactions, which in a sense cancel each other out. The billiard ball model is the starting point for the kinetic theory of gases. For example, in a container full of very many particles (atoms, molecules, whatever) whizzing around at all sorts of speeds, bumping into one another and into the walls of the container, we have one single property that is stable with respect to type of gas, shape of the container, nature of the walls and so on. That property is temperature. We call it an emergent property because we wouldn't have expected it just from looking at individual particles and we can't measure it for any single particle. Unfortunately emergence is almost as hard to define as complexity. Emergence is one of the hallmarks of complexity. Out of a set of simple processes comes something unexpected on a global scale. We shall see many examples throughout the course of this book. In the end we shall have a feel for what it is, but still perhaps not a formal definition.

Temperature is a good example of an emergent property. What is it? We can measure it on a macroscale and we have all sorts of physics dependent upon it. But at a molecular scale it disappears. We have only many molecules or other particles (nuclei in a plasma, etc.) with a wide variety of velocities (and consequently energies). It is only the collective distribution that gives rise to the single scalar quantity temperature for the assembly.[3]

What is true of billiard balls is true of other systems too. Whether it be gas molecules in a jar, stars in a galaxy, cells in our bodies or people living in a town, it is the interactions that make each system complex. Traditionally each of these systems has been studied separately. Gases have been in the domain of thermodynamics, cells in biology, and towns in sociology. Only recently have scientists begun to appreciate that, despite their obvious differences, all these systems share important properties in common. Is the whole greater than the sum of its parts?

A PARADIGM SHIFT IN SCIENCE?

As the science of putting things together, the study of complexity is a radical departure from the traditional way in which science has been done. Traditional science uses what is known as the *reductionist paradigm*. The idea is that you can understand the world by breaking things down into their components. Complexity is exciting because, while the components may be trivial, their *interaction* often generates wild and unpredictable behaviour.

For instance, if you want to understand the relationship between a forest and the environment, then you look at plant physiology. You study the way plants pump water and you study the biochemical pathways involved in respiration and photosynthesis. You perform experiments in which you measure the exact rates of water

9

transpiration under different temperatures and air pressures. This is the reductionist paradigm. It is a way of doing science that has served us well for hundreds of years. It has led to countless discoveries and has led to the technical wonders that we are so used to today.

So, at the end of all your experiments on plant physiology, you know exactly how a forest works. Or do you? You certainly know a lot about the workings of an individual plant, and that can tell you a lot about the nature of forests. It tells you where plants can grow and where they cannot. But does it tell you everything?

Fortunately for field ecologists, the answer is a resounding no. For the truth is that a lot of features in a forest are the results of interactions. Studying an individual plant does not tell you how it is going to interact with other plants. Trees in a forest may seem stately and quiet. But in reality a forest is seething with competitive and symbiotic interactions. They have a telling effect. Plants compete with one another for space, for light, for water and for nutrients. They disperse seeds. They provide homes for birds and animals. They are attacked by insects and other herbivores. They create leaf litter, which develops into soils and provides an environment for millions of insects and micro-organisms. Without all the richness that these interactions bring, a forest would be little more than a sea of potential telegraph poles.

The reductionist approach is to reduce everything to the absolute minimal level. So if we could understand a forest by studying plant physiology, why not understand plants by studying their cells? Why not understand cells by studying biochemistry? Why not understand biochemistry by studying nuclear physics? The ultimate expression of this approach is the quest for a unified field theory. Einstein spent the last years of his life searching for a way to link gravity and the electromagnetic field with the strong and weak nuclear forces. The quest continues today, with many successes and failures along the way.

But if the quest for a unified field theory ultimately succeeds, if someone does create a 'theory of everything', will that be the end of science? Should we all pack up our bags and go home? This is a far from trivial question. Paul Dirac, predictor of anti-matter, leading quantum theorist and Nobel laureate, asserted that chemistry had become a branch of pure mathematics! Can we start from these basic rules of fundamental science and predict how a large system will behave? For a long time the answer seemed to be yes, prompting all sorts of arguments about the existence of free will and other matters of metaphysics. We now know that it is not always true. Even using huge computers? In fact the amount of computation required grows so rapidly that, for even simple chaotic systems, prediction on large time scales is just not feasible, ever.[4] There are other complications to do with quantum mechanics, but we will not dwell on them here.

Simulation of a complex system may be very useful practically. Boeing design aircraft on computers rather than using wind-tunnel models, on grounds of speed and cost. But simulation doesn't help us *conceptually*, in understanding the rules of behaviour at the higher level. So are there methods to develop that give us the rules of the emergent system? Just as with our attempts to understand the nature of a forest, there are many things that the lower-level phenomena cannot explain. In fact there is a subtle computational argument for believing this to be the case, which we shall consider in our discussion of the fundamentals of computation.

The most intriguing example of an emergent phenomenon is surely human consciousness. The human brain has many nerve cells, but the number of individual types is very limited, certainly less than 100 but not

11

much more than ten. The dynamics of a neuron are no deep mystery and the way neurons adapt (learn) is still not completely solved, but the broad outline seems to be there. So where or what is the human mind? This deepest of philosophical questions is still with us and not obviously soluble through ever deeper study of the behaviour of nerve cells. It is somehow a phenomenon which emerges from the grand ensemble.

To truly understand the world it is not enough to take it apart. We also have to learn how to put it back together. The reductionist approach has been an enormously powerful tool for science, and it will continue to be so. However, we cannot blind ourselves to the need to understand just how and why the whole is so often greater than the sum of its parts. Nor is this need merely curiosity. The ultimate complex system that we have to deal with is the world we live in. If humankind is to survive and prosper then we have to learn how to manage this complex system of which we are a part.

CHAPTER 2

COMPUTATION

Before we examine the links and analogies between science and computation we need to ask what computation really is. Is it just what computers do? Hardly, since computers vary dramatically from month to month and we can do most things with a pencil and paper (maybe not so fast!) that a computer can do.

The computers we use in our homes and offices are usually derived from the model put forward by John von Neumann. In this model computers consist of a memory and a processing unit. The processing unit fetches data from the memory, carries out arithmetical or logical operations and writes new data back. The instructions for the processing unit constitute the program. Additional operations include control operations which allow the processing unit to repeat, or iterate, a set of operations or to select different instructions according to some data value.

This is a very functional description. At a more abstract level what our computer does is to map one set of numbers (integers) to another. We say that the output is a function of the input; alternatively, a *mapping* from the input to the output. For finite data sets for input and output we can make a list of which output goes with which input. This could get very cumbersome, but it sets an upper limit to the size of a program and there is no reason why we shouldn't be able to compile this list for any given case. However, when the range of possibilities becomes infinite, encompassing all the integers, it turns out that not all such mappings are computable. We could say that 'computable' just means 'can be computed on a PC or a Mac'. But computers come and go very quickly, so what we need is not an example of a present-day computer but some more abstract general model of a computer. Enter the Turing Machine.

TURING MACHINES: THE ESSENCE OF COMPUTATION

Alan Turing proposed a definition of 'computable' based on a simple abstract computer which bears his name and is now the building block of elementary courses in computer science. The Turing Machine (or TM for short) is a model of a computer reduced to the absolute basics. In Turing's model the computer consists of two elements: a tape and a controller. The tape contains three sorts of symbols, blanks, zeros or ones. These symbols can represent information in various ways. For instance, a string of six ones might represent the number 6. The controller is a device that performs various operations on the tape. It can move the tape one step backward or forward and it can read or write one symbol at a time. To make it work the controller has two other features. The first is its program, which determines its behaviour. The second is its 'state', which is somewhat like a person's mood in that it affects the way the controller behaves.

So, given a tape with some symbols on it and a controller with a number of states, the TM will go into action. The program consists of rules that tell the machine what do to for each combination of state and

14

tape symbol that it encounters. The following list sum-
marises the rules that make up a simple program.

STATE	TAPE	ACTION
0	0	advance the tape forward
0	1	change state to 1, advance the tape forward
1	1	advance the tape forward
1	0	write 1, change state to 0, advance the tape forward

Suppose that the TM starts in state 0 and that we
feed into it a tape with the following string of symbols
on it: 00011110000. As the machine reads this tape it
behaves as shown below.

INITIAL STATE	INPUT SYMBOL	OUTPUT SYMBOL	FINAL STATE
0	0	0	0
0	0	0	0
0	0	0	0
0	1	1	1
1	1	1	1
1	1	1	1
1	1	1	1
1	0	1	0
0	0	0	0
0	0	0	0
0	0	0	0

The net result is that the machine has increased the
number of 1's on the tape by one. That is, it has changed
the number 4 to 5. So the function of this machine is
to add one to any number fed to it. Now this may not
exactly be a very profound result, but this is just a simple
example. The point is that it *is* a well-defined function.
What Turing showed was that, simple though it may be,

15

Figure 2.1 The operation of a Universal Turing Machine. One tape holds data, the other a program.

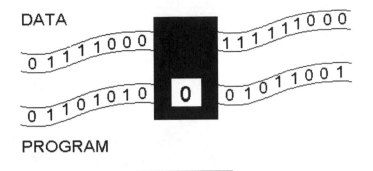

DATA

PROGRAM

we can design machines like this to carry out *any* arithmetic computable function we can think of.

An important limitation of the TM is that each machine performs one, and only one, function. To cope with this problem Turing introduced a second tape drive. This new device, called a Universal Turing Machine (or UTM for short), reads its program from the second tape. Turing showed that the UTM is capable of emulating any simple TM. This property, called *universal computation*, is the essence of any modern computer.

Now, as a device the UTM is hopelessly impractical. To try to solve real problems with it would be extremely slow and awkward. Nevertheless it does encapsulate the main features of any modern computer. The UTM provides a useful model for asking general questions about computing.

Before this *digital* model of computing, based around mapping sets of integers onto each other, people talked of *analogue* computing. Lots of physical devices, such as thermostats, are analogue computers. The old black vinyl records are an analogue recording medium, unlike CDs

16

which are digital. On a vinyl record the sound signal is related directly to the changes in shape of the record groove, the general principle of analogue systems. Some hi-fi enthusiasts maintain that analogue records are better than digital, but, if that were true in the early days of digital, the difference is fast disappearing. One of the beautiful features of the Turing model of computation is that *everything* in the classical world is computable digitally.

THE HALTING PROBLEM

An important question in computing is whether particular problems can be solved. One way to approach this question is to ask how long it will take. If a TM can solve a particular problem, it will stop when it has finished. It will stop after a finite number of steps.

But what if the TM cannot solve the problem? Will it ever stop? This question—called the *halting problem*—turns out to be unanswerable for any arbitrary controller. There is no function that a Turing Machine can represent that will determine whether an arbitrary Turing machine program will terminate. But any function represented by a finite controller that does stop is defined as a computable function. The unpredictability of emergence arises from the impossibility of determining if a Turing Machine will halt.

Turing's machine is just one possible procedure for carrying out a functional mapping: could there be better ones? Lengthy research over the subsequent decades has not found an alternative. In fact, Alonzo Church, another prominent figure in the theory of computation and Turing's PhD supervisor at Princeton, asserted that *effective calculability* is independent of formalisation, a thesis which still stands today.

In modelling science by mathematics, we implicitly

assume that changes in state in the natural world are all computable. So far, there is no counter example, or likely counter example. Yet it is still possible to find people with an intuitive (only) feeling that there are things a brain does which do indeed lie outside the class of computable functions. But the brain is steadily yielding its secrets. We now know much more about biological neural systems and can also model neural computation quite successfully on computers. From the classical standpoint we see no reason to doubt that the computational strategies of the brain will eventually be fully understood.

Computational hierarchies

Complicated computer programs are difficult to write and, since the early days of the development of programming languages, modular structure has been a key tool in managing complexity. The next step beyond this was to encapsulate both data and program into independent modules, or objects. This new paradigm swept through computer programming during the 1980s. As computers have entered more and more of our lives, as hardware improves rapidly, the need for robust software gets ever greater. Yet the design difficulties have proved so great that people talk regularly of a software crisis. Many see object-oriented methods as a major step forward. An object is a collection of data and the operations that access and manipulate that data. By hiding the data, and allowing access to it only through a limited set of special methods, the complexity of software development can be broken down into manageable and provably viable pieces.

Natural complex systems exhibit modularity almost everywhere, from the building blocks of evolution to the structure of the human brain. Thus the object-oriented approach has the right feel about it. But alongside the

18

development of object-oriented methods the networking of computers has mushroomed, bringing with it the need for objects to be accessible across networks and for objects to be self-describing. Soon we will have the option where one object may go searching on computer networks for other objects about which it knows nothing in advance. The stage is set for emergent computation on computer networks! We do not yet know what will happen.

A PHOENIX RISING FROM THE ASHES

'We are on the threshold of a new era in computer architecture.' So began an excellent computing text, *Highly Parallel Computing*, by Almasi and Gottlieb, published in 1989. At that time, with defence money pouring in, massively parallel computers were the cutting edge of supercomputing. Machines such as the *Touchstone Delta* (Intel) at Caltech or the *Connection Machines* produced out of Boston by Thinking Machines Corporation vied with each other for supremacy in a range of benchmark mathematical, scientific and engineering problems. Other companies such as INMOS and Texas Instruments were building plug and play parallel computing chips. Parallel computing was the way forward, or so it seemed.

The obvious advantage of parallel computers is that they can perform many tasks at the same time. Imagine a supermarket checkout during peak hour shopping. A shop with just one checkout is like a computer with a single processor. During busy periods such a shop is liable to have long queues of waiting customers. Supermarkets solve the problem by adding more checkouts. As a group, twenty checkouts can process customers twenty times as fast as one can. Likewise, a computer such as the Connection Machine CM2 with (say) 65 536 processors can carry out a lot of similar calculations thousands of times faster than a single processor.

But parallel computers turned out to be hard to program. Getting messages between processors and making sure that all processors are fully occupied all of the time is far from easy. It's not only difficult, it's difficult to automate, making parallel programming a rather specialist art. Worse still, parallel computers had to survive in an extremely competitive computing market. By 1996 the situation had changed radically from 1989. Almost every computer manufacturer dependent on parallel computing was in trouble: Thinking Machines Corporation had been dismembered after serious financial difficulties; Cray Research had collapsed; Intel had ceased work on its range of parallel computers; Convex went under. (Intel of course had a booming semiconductor business.) Two things had happened: firstly, the collapse of the Iron Curtain dramatically cut defence funding and the Californian freeways once full of cars into Silicon Valley were almost overnight congested no longer; secondly, the growth in microprocessor performance was greater than anybody had expected. Make no mistake, the most powerful machines on the planet are still parallel: but the mass market opportunities are gone. In their place we have distributed computing with independent machines loosely coupled by computer networks.

One method of distributing computation is to carve up a large problem into many small, separate cases and to assign each case to a different machine. Computer networks such as the Internet serve as a parallel computer, but one in which communication between the processors is very poor. For instance, simulations of the world's weather divide the earth's surface up into many cells and assign each cell to a separate processor. However, to simulate the circulation of wind and water, those cells have to exchange information with neighbouring cells constantly.

Large machines with many interconnected processors on special high speed communication frameworks are much better suited to problems where the different cases have to exchange information, and problems such as this are still best handled by the big (very expensive) purpose-built parallel machines. The leading edge of high performance computing is still parallel.

Other problems do not require this tight coupling, and heterogeneous computing networks are viable. Updating all the records in a large database, for example, would just involve each processor taking its share of records. Communication between the processors would not be necessary.

COMPUTATIONAL COMPLEXITY

We now have a simple model for universal computation. Next we need to look at what constitutes complexity for a computer and see how this might relate to complex systems in the natural world.

Hard problems

In Douglas Adams' novel *The Hitchhiker's Guide to the Galaxy,* scientists ask the computer Deep Thought whether it can find the solution to the ultimate question of life, the universe and everything. The computer replies that yes, it can, but the solution will take some time— eight million years. As outrageous as this period may seem, it is not difficult to find problems that would take much longer than eight million years to solve. The question of time to solve a problem is relevant whenever we try to scale up from a small problem to a large one. Most problems can be solved quite easily for small cases,

21

but what happens as we increase the number and complexity of cases involved?

What *makes* a problem hard to solve on a computer? The complexity of a computational problem has been thought of as being composed of the time it would take to solve the problem and the working space needed to store all the intermediate results. In computer science the first of these, the time taken as a problem increases in size, often receives the most attention. We could use some measure of the number of steps it takes to solve a problem as a measure of the time it would take. Since computers are getting faster all the time, it would make little sense to use an actual time, although in principle we might standardise on some particular machine. But then we would have to rely on standards museums to keep such machines alive and working, which does not seem a very practical idea. Instead, what we do is to determine how the number of steps grows in size as the size of the problem measured in some way increases. The time taken or the number of steps required depends of course on how we go about the solution—the *algorithm* we adopt. There are exact and approximate algorithms and also good and bad ones.

One of the lessons from natural computation (discussed later) is the emergence of fast approximate solutions. Needless to say, these are of potentially great interest in many application areas. To make this a little easier to grasp, let's look at a couple of examples. An easy problem is a simple search. Suppose we have a number of identical containers in which we have to search for something. If it takes the same time to search each container, then the time taken will increase linearly with the number of containers: if we double the number of containers it will take twice as long and so on.

A somewhat harder problem is to determine which two of a set of cities are the closest together. To create a table of distances between cities we need to do a calculation for every possible pair. For each new city that we add to the list, we have to calculate its distance from every other city. For 5 cities, we need to calculate the distance from the first city to all 4 others, the distance from the second city to the 3 remaining cities, and so on. So the total number of calculations is 4 + 3 + 2 + 1, giving a total of 10 cases to calculate. For 100 cities this number grows to 4500 cases. Any home computer could still knock this over in no time flat. In fact the size of the problem increases with the square of the number of cities. We say that the growth is *polynomial* in the number of cities.[5]

A well-known hard problem is the travelling salesman (this should now be travelling salesperson, but we need to stick with historical usage). Suppose that our salesman needs to visit each of a set of cities. Then, to save time and money, he needs to know the shortest route that visits each of the cities just once. Unlike the simple table, if we now add a new city to our list then we have to check its effect, not on each other city, but on every possible tour of those cities. So each new city multiplies the number of possible routes. For 5 cities there are only 12 cases to examine (the length of the path is independent of the starting point and direction). However, the size of the problem escalates at an alarming rate. The figures in Table 2.1 show just how rapidly the complexity of the problem goes off the scale.

Suppose that we have a supercomputer capable of checking a billion cases per second. Then just 28 cities are enough for our supercomputer to take longer than the age of the universe to solve the problem. This is not reasonable at all! In fact, the number of routes grows *exponentially* with the number of cities.

Computer scientists make a precise distinction between classes of problems on the basis of the rate at

Table 2.1 Number of tours as a function of number of cities

Cities	Tours
3	1
4	3
5	12
6	60
7	360
8	2520
9	20 160
10	181 440
15	4 358 914
20	10 000 000 000 000

which growth occurs. The Travelling Salesman Problem (TSP for short) belongs to a class of problems known as 'NP complete' (NP is short for 'Non-deterministic Polynomial'). These are problems that quickly grow unmanageable as their size increases. In many of these problems the complexity stems from permutations and combinations of simple elements. In the TSP, for instance, each path can be written as a string of cities ABCD . . . or whatever. So the number of cases depends on the number of ways in which we can order a string of symbols.

This abstract description reflects an important feature of computer science: many apparently different problems are equivalent to one another. So the solution to one problem typically allows us to solve many related problems. For instance, some similar problems include trying to find the shortest route through a maze of streets or trying to use the least amount of wire when installing electricity in a house. At a more general level, the TSP is one special case of the so-called scheduling problem. Other examples of scheduling include arranging classes in a school, timing traffic lights, land-use planning, and programming work on machines in a factory.

24

SEARCHING AND SOLVING IN THE NATURAL WORLD

Animals spend most of their lives searching for food, mates and shelter, but searching is also used as a computational metaphor, where we think of searching in a solution space. Unlike the TSP, search problems are often solvable in a reasonable time. First of all, we do not need to test every possibility to solve the problem. We can stop as soon as a solution is found. Secondly, we can often discard many cases without checking them individually. For instance, suppose that we are searching a medical database for people satisfying a set of selection criteria. Then if a person fails the first criterion we can immediately discard that record; there is no need to check all the other criteria. So what do we do about problems like the TSP that need millions of years to solve? City planners simply cannot wait around for centuries to solve the problems of choked urban roadways. And to say that a problem is insoluble helps no one. Real world problems need real world solutions, fast, and roughly but not exactly correct.

We can begin to get somewhere if we note two points. Firstly, people usually need a good answer, but not necessarily the best. Secondly, as in life it pays to compute smarter, not harder. Earlier arguments about the TSP concern 'brute-force' methods in which the computer laboriously checks all possibilities, one after another. Brute-force methods are not always feasible. How *do* travelling salesmen plan their routes? One approach is to use a technique that might be termed 'divide and rule'. One practical aspect of travelling is that you have to stay somewhere. It is more convenient to stay several nights in one hotel than to be constantly moving. So travellers usually base themselves in major towns and make a series of short side trips to nearby

25

towns. So if you had to visit 20 towns you might divide them into (say) five groups of four towns each, with a central base in each group.

The problem of planning the shortest route thus reduces to finding the shortest route around the five base towns. So instead of a vast number of cases the traveller has only twelve cases to consider—a task that can easily be done by hand on the back of an envelope. The basic idea embodied in the divide and rule approach is to break a big problem down into small ones for which a brute-force approach is viable. This is yet another example of a computational hierarchy.

One way of visualising the business of solving complex problems is to think of a *solution* landscape in which all possible solutions to a problem are laid out on a map. Solutions that lie near one another are similar in some respect. Also, if we suppose that elevation represents the quality of a solution (e.g. the travelling time in the TSP) then we can imagine solutions as looking like a landscape with hills and valleys. So the task of examining one potential solution after another is like wandering around in the landscape where the desired solution is the highest hill (or deepest valley, depending on the nature of the problem). This image of a solution landscape derives from problems in which solutions can be expressed as numeric values of parameters. For instance, efficient production in a factory might depend on a number of factors such as length of production runs, stores of raw material and so on. If we treat the values of these parameters as akin to latitude and longitude then we can literally plot values in a solution space.

Supposing that we can structure our search as a solution space, how does that help us? Well, first of all it helps us to reduce the search time. For instance, if we start anywhere in the landscape we can immediately find a better solution by wandering about. How does a blind mountaineer find the top of a mountain? The answer of course is by going uphill. So, starting anywhere, we simply go uphill until we cannot go any higher. For simple problems this approach (called 'gradient ascent' or 'hill-climbing') is a good way to find the best solution.

In general, however, a solution space may be a complex landscape containing many hills and valleys. The problem is that hill-climbing yields only a local optimum. There is a risk of becoming trapped in the foothills and missing the mountain altogether. What we need is a method of surveying all the major hills in the landscape.

This is an area where computing can learn from nature. Nature has evolved many ways to solve complex problems. These have provided inspiration for many algorithms, or methods for solving problems. Take an ant colony. The colony needs food to survive and has (literally) to search the landscape to find food. As we all know, the colony sends out scouts to locate food. Note that it sends out many scouts, not just one. The scouts have no knowledge of the landscape, nor do they know which areas are more likely to hold food than others. They have no intelligence, so they cannot plan a systematic search to take in every corner of the local area. So what do they do? Simple. They just wander around at random until they find something. That something may be food, or it may be some clue, such as a scent, or a pheromone trail left by other ants.

When they do find something the ants head off to the nest, leaving a trail so that other ants can follow the same route back to the prize. This strategy is far from perfect. Ants do miss a lot. They often walk straight past food without detecting it. However, that does not matter. The goal is not to find every scrap of food but to find enough food. What matters is retrieving the food

27

that they do find quickly and efficiently before someone else does.

Every species has its own foraging strategy. Contrast the ants with (say) an eagle. Unlike an ant the eagle literally does have an overview of its entire environment. However, its food, such as mice, rabbits or lizards, tends to be rare and tries to evade detection. So the eagle's strategy is to hover in the sky scanning the ground for any hint of movement. The chances of finding a rabbit under any one bush is small. But by scanning the whole landscape the eagle is able to spot any error made by any rabbit in the entire area.

Evolution itself can be seen as a search technique. One way of looking at evolution is that it is a grand experiment—a search in which nature is constantly experimenting with new designs for making living things. Good solutions are animals that survive and breed to produce offspring. Two things to note about evolution and natural selection are, first, that it throws up not one solution but lots of solutions and, second, that its starting point is always a previous solution that was successful. This appreciation of evolution led John Holland to develop a computational version of evolution. This technique—the genetic algorithm—mimics the above process to solve complex problems.

Evolution, brains (neural systems) and the immune system have all spawned computational methods. We consider some of these methods now.

SHORT PROGRAMS FOR ORDERED SYSTEMS

Staring into the fire, in the days before television, we would see the flames shape themselves into familiar objects such as faces. This idea of patterns in random systems appears in many places: there is even a psychological test based upon what people perceive in ink blots. This quirk of human nature conceals two very important ideas. The first is the nature of order and randomness. The second is the search for order which drives our brains.

If you look at an active television not tuned to a station you see a random pattern, so-called white noise. Now, just as in the fire or an ink blot we might be able to interpret this rapidly changing pattern as being like some natural image; in practice we don't see anything really approaching the natural world. If we think of the television picture as being made up of an array of *pixels*, each pixel having a particular colour, then from the number of pixels and number of colours we can calculate the maximum possible number of television pictures.[6] It's not difficult to imagine that we would never in our lifetime see a real picture and thus we can accept intuitively that the proportion of random pictures is very high. In fact randomness is a tricky concept and it takes some sophisticated mathematics to make it precise.

We have seen the notion of the time it takes to solve a problem as a measure of complexity. This is a fairly old, traditional notion. However, in the 1960s the great Russian mathematician Kolmogorov, founder of modern statistical theory, came up with another complexity idea. As with many breakthroughs in science more than one person may be very close, and in fact what is now called Kolmogorov complexity was essentially invented simultaneously by Chaitin in the United States, and Solomonoff, also in Russia.

The idea is that any object can be described by a computer program. The length of that program is the complexity of the object. So if we compare a picture of a human face with a chequerboard pattern, the face is much the more complex: the chequerboard pattern can be described by a very simple formula. (We might compare abstract impressionist paintings along these lines: Mark Rothko's vast canvases have a few more or less uniform colours in large patches; Jackson Pollock's equally vast canvases are extremely complex with much tiny detail, and are *very* difficult to describe.) It might seem at first sight that this would be a very uncertain quantity, strongly dependent upon the computer and the language used to program it. Fortunately this is not the case, when the programs are large enough. We can see this simply by taking some common language for the description and just adding a translator at the beginning for any other language. The translator will be of a finite size and fixed independently of the problem.

Many complex systems display elaborate and complicated surface phenomena, although beneath the surface very simple rules are at work. Thus they may be modelled with a short program. At first Kolmogorov complexity seemed like a route to the measurement of complexity itself, but since the most random objects have the highest values it does not fully capture the notion of complexity. Murray Gell-Mann, Nobel Laureate and co-founder of the Santa Fe Institute, expresses it neatly: surface complexity with underlying simplicity. As a measure of complexity we need a combination of short program and complex output.

Noise, entropy and information

Putting names to faces can often be difficult: not long ago, at a conference on information theory at Brighton in England, somebody casually registered under the name of Claude Shannon. The conference got under way without anyone taking particular notice of this registrant. Then whispers started that this was in fact *the* Claude Shannon, who wrote the textbook for communication theory and demonstrated that exact communication was possible even in the presence of *noise*. (Not all pioneers are attention seeking!) Although in principle we might measure something with arbitrary accuracy, in practice noise prevents us from doing so. Shannon, however, showed that it was possible to define the information capacity of a channel, even if this channel was noisy. Furthermore, the channel could transmit information encoded in such a way that, provided this capacity was not exceeded, the signal could be reconstructed perfectly, at the other end.

Thus for present purposes we can just ignore the arbitrary accuracy of physical measurements and just assume that they are discrete, with the interval between gradations being determined by noise.[7] But in fact there is a more fundamental result, established only in the early 1980s, which relates computation directly to energy.

Thermodynamic cost of computation

For some time, debate has raged over whether or not computation requires energy. True, the Turing Machine is a hypothetical machine, with zero weight tapes and frictionless movement. So just how far could we push the limits of computation as computers get better and better? Is there some inevitable loss of energy as heat, as with all machines? Perhaps not: Landauer and Bennett at IBM established that computation in terms of changing information from one form to another does not require any energy, and we should be able to build a perfect computer.

But the *destruction* of information takes energy. In very simple terms, if we carry out a very complex calculation but save all the intermediate steps, then we can carry out the calculation backwards: it is reversible. But as soon as we destroy the intermediate working, then reversibility is lost and at this point energy is expended.

31

Traditional statistical thermodynamics has always recognised the intricate relationship between order and energy. Breaking up ordered systems, as in the change from a solid to a gas, involves an increase in *entropy*, or disorder. Thus going from solid to gas requires less energy than that required simply to break the bonds holding the solid together. These ideas go back to Gibbs in the ninteenth century. But a much newer result is that the destruction of algorithmic information also has a penalty.

In what sense can we think of the world as computation? The Turing model focuses on inputs and outputs, but a physical or biological system is analogue. Even if we ignore quantum mechanics, there is still a difficulty in moving from simulating physics on computers to physics being computation, at least in the Turing sense. Some biological processes are obviously rule-based and these map readily to the Turing Machine model. But an alternative is to think not in terms of algorithms but in terms of the creation and destruction of information. Now we can put a precise limit on the cost of destroying a single bit of information, kT, where k is Boltzmann's constant and T the absolute temperature, exactly the same energy as per degree of freedom of a molecule in a gas.

How efficient, then, are today's computers? So far they are nowhere near the above theoretical limits. Recent work at the University of Cambridge has estimated precisely the energy used in neural computation in insects. It too is a factor of millions above theory. Biological sensory systems have evolved to get very close indeed to the physical limits: many animals can detect single photons and some have hearing that is limited only by thermal noise in the membranes of the ear. No creature in the universe could do better. Computers, it seems, still have a long way to go.

And so what of quantum mechanics?

The complexity of the world of classical physics is rich enough. But then we come to quantum mechanics, which replaced Newtonian mechanics at the beginning of the twentieth century. David Deutsch, a physicist at the University of Oxford, created a quantum analogue of the Turing Machine. It can perform some calculations faster than a classical machine can. Unlike a conventional Turing machine, the Deutsch Machine has an array of tapes that are blended when the machine stops. Theoretically this is a very interesting field. There is active research in how to build a quantum computer and research into the sorts of problems that might be solved faster by it.

David Deutsch has shown that the computation of the real world needs a new type of quantum Turing Machine, while Roger Penrose has suggested that quantum phenomena may be fundamental to the working of the human brain. Although these ideas are profoundly original, there are still many classical avenues to explore before we give up on the mysteries of the human mind.

Later we will talk about local rules in cellular automata, neural networks and self-organising systems and how local rules produce interesting global emergent phenomena. But the message of quantum mechanics is that the world really is non-local. In the next section we describe the remarkable experiments that demonstrate fundamentally non-local phenomena. Yet we do not know if these have any macroscopic effects.

Non-locality

The two twentieth-century revolutions in physics—relativity and quantum mechanics—have in common some very unintuitive phenomena. In quantum mechanics one of the strangest is *non-locality*, a sort of instantaneous action at a distance. Very precise experiments have tested this notion. Quantum mechanics gives us two complementary perspectives on light. For example, we can view it as a wave motion or as a stream of particles called photons. Each photon has a spin, just like a top or the rotation of the ear. For complicated reasons, we can't give a very precise value to the size of a photon. Suffice it to say that photons are *very* tiny—much smaller than the smallest atom. Yet the most recent experiments involve matching spins of single photons over a distance of several metres! (See page 35.)

There isn't any causal link, though, since this would violate the principles of relativity, where no communication can occur faster than the speed of light. Since effects such as this are so unlike anything that we encounter in our daily lives, that we encounter with our senses, we resort to stories and analogies to picture what is going on.

To understand non-locality, let us consider a simple story, which follows Roger Penrose's Quintessential Trinket argument. Imagine a gift shop situated in a remote part of the galaxy. It is important that it be a long way away, since we want to be sure that no secret messages, travelling at the speed of light, could arrive in a short space of time. One of its gifts is a little goblet set packed in pairs, one silver, one gold. The shop will dispatch one of each pair to one of any pair of friends, pre-wrapped so that even the people in the shop don't know who gets the gold and who the silver, except that there is one of each. So if one goblet is sent to you and the other to a friend light years away from you, it will not be until you open it that you know which each of you got.

There is nothing difficult about this: the goblet in *your* package is either silver or gold, but which it is just not known until the package is opened. The same goes for your friend.

But there is another gift for each of you: a Quint-essential Trinket, a little toy in the form of a dodecahedron (a solid object made up of twelve faces each with five sides). On each vertex is a button: pressing the button may do nothing or it may ring a bell and cause the trinket to vanish in a puff of smoke. Now, the magic here is that the shop guarantees that some combinations of button presses by you and your long-distance friend will cause bells to ring; other combinations will not.

It is impossible to find in physics a solution to how that information could be imbedded in the trinket itself. There must be some information flow (impossible because of the distance) or some very spooky properties of the interior of the trinket. These properties are exactly what quantum mechanics predicts. But, whereas we have an explanation for the pairing up of the goblets (pre-packed and labelled), there is absolutely no explanation in classical terms for the properties of the trinket.

This seemingly instantaneous action at a distance was first raised as a paradox by Albert Einstein, Boris Podolsky and Nathan Rosen in mid-1930s. A photon has associated with it a *spin*, which may have two values, up or down (almost as if it could turn in either direction). The essence of the paradox is this: when two photons are created from (say) the collision of a particle and its anti-particle equivalent, their spins are created equal and opposite. But which has which spin is not decided. They can travel to the opposite ends of the universe and either photon can assume either spin. The crunch is that when one photon's spin is determined the other, at the other end of the universe, is also determined: no communication takes place (since photons travel at the speed of light this would be rather difficult anyway). Somewhat later, John Bell showed that the results could not be the result of some hidden mechanism or concealed information buried within the trinkets.

This 'action at a distance' that quantum mechanics gives us is philosophically very confusing. Although these effects are instantaneous across the universe and seem to defy relativity, which limits the speed of travel to that of light, they cannot be used to transmit information. The universe is somehow entangled in past histories of its constituent matter. We have yet to find any ways in which this manifests itself at a macroscopic level.

So we find ourselves in an interesting situation. At the beginning of the century the simple classical picture of the world gave way to quantum theory. The Heisenberg Uncertainty Principle destroyed, forever, the idea that we might have perfect knowledge of any physical system. The non-local properties gave very unintuitive predictions about observations made at a distance.

At the close of the century we have seen the chaos revolution and the realisation that the classical physics of Newton had nothing like the strict determinism we always imagined it had.

CHAPTER 3

MILITARY AND DEMOCRATIC ROUTES TO COMPLEXITY

When we watch a military parade, we see squads of soldiers marching in perfect step. Their walking is *synchronised*. We don't need to track every soldier individually: we need only to know where the squad is and what direction and speed it's going in. With just this small amount of information, we can determine the movement of every soldier.

Soldiers are trained to march. But many natural systems show synchronisation like this without any external influence. On the other hand, animals such as ants cooperate, producing complex behaviour for the colony, but do not have this military precision.

MARCHING IN STEP

The laser is a modern embodiment of the ray gun of science fiction. But despite Ronald Reagan's Star Wars efforts, the laser's main uses are in technology, not war. The laser is now a common household device: all CD players, for example, use lasers to read compact discs.

Lasers have come a long way from science fiction. We have also come a long way from the theoretical turbulence the laser created. According to statistical thermodynamics, the atomic or molecular particles of a gas have a wide range of energies. This distribution tails off at higher values and is referred to as the Maxwell distribution. This distribution characterises *thermal equilibrium* brought about by collisions amongst gas molecules. Temperature is defined as the average molecular kinetic energy. In a laser, just before firing, the gas is far from such an equilibrium. The advent of the laser thus introduced a new system, not in equilibrium and with a vast distortion of the energy distribution.

The activity of a laser occurs in two stages: in the first (pumping) stage, light is applied to the laser gas, injecting energy; in the second stage the energy is released in a coherent pulse. In the pumping phase (we shall gloss over the technical details) gas molecules are raised to an excited energy state as they absorb light. In traditional theory they would shed this energy by collisions, the energy distribution would even out according to the Maxwell distribution, and the overall temperature would rise. But for a variety of reasons, amongst which is the rate at which energy is being pumped and trapped in the system, this does not happen and we end up with a population inversion where there are far more molecules in higher energy states than in lower—a lot more. The right trigger can now release all of this excitation energy in a single coherent pulse, the laser beam.

One of the prominent early theorists of the laser, Hermann Haken, is also one of the great unsung heroes of complexity. Haken's analysis led him to formulate the 'slaving principle', which he then proceeded to apply to many other systems, developing a field he called *Synergetics*, the science of *self-organising* systems. In fact Haken has hosted a series of conferences on Synergetics going back nearly two decades! The mathematics of Synergetics is quite difficult but we can explain the two core ideas. A complex dynamical system has many variables, but under the right circumstances some variables will enslave the others, just as the drill sergeant keeps the soldiers marching together. The system acts in unison under just these dominant variables. But for this to happen, noise—or in Haken's terms, fluctuation—is essential to give the enslaved modes enough plasticity to get into step. There are many simple systems that exhibit slaving, without us needing to look at systems that are obviously diverse in number and type of agents, such as ecologies.

Something as simple as a boiling liquid exhibits a synergetic phenomenon—Benard cell convection. The liquid self-organises into columns of hexagonal cells: in some columns liquid travels up from the bottom of the vessel, while in adjacent columns liquid travels down. Just think how extraordinary this is: there are no inherent boundaries in the liquid and these columns have formed spontaneously. Furthermore, they have a clear geometrical shape, again something in no sense obvious from the initial setup.

Marching in step is an example of rhythmical, coupled movement. Periodic motion, from pendulums to planets, is something we learn about in early childhood. Before the advent of quartz watches it was the fundamental component of a timepiece. Quartz devices, of course, are still oscillators, but the oscillations are hidden down at an atomic scale. A simple oscillation has just three properties: the frequency with which it returns to the same point in the cycle; the amplitude, which is a measure of the power in the oscillation; and the phase, the point at which the cycles begin.

We see interaction of oscillators in many places in everyday life. In an early differentiation of the sexes, Mary has her two small children Jack and Jill on the swings in the park. Jill rocks gently back and forth, but Mary gives Jack's swing just a little nudge at the high point and the swing gets bigger and bigger. But Jill does not want to be left out. Each time Jack passes she tries to grab him. As she touches him she slows him down and he speeds her up. Before long, both are swinging together. Later on we shall show how this situation of the swings together is an *attractor* in the joint dynamics of two systems. Part of the idea of an attractor is that it is easy to get in and hard to get out. Hence, often, only small amounts of energy are needed to bring about synchronisation of behaviour.

Several simple oscillators that can couple with each other in some way display an interesting effect: even if the frequencies are not all exactly the same and even if their phases are random, eventually they will lock together, with the same frequency and phase of oscillation. This remarkable effect was first observed by Christian Huygens, the inventor of the wave theory of light, in the synchronisation of clocks on a wall. They might start off all reading different times, some running fast, others slow, yet after a while they will all end up going at exactly the same rate. Adjust them to all read the same time and they will stay that way, although they may all be running at the (same) wrong rate. For this *phase-locking* to occur the clocks must be able to link up in some way: by putting them against a wall, their mechanisms transmit vibrations into the walls which serve to couple them together.

Phase-locking is a common problem in bodily activity. For instance, a common childhood game is to try to pat your head and rub your stomach at the same time: it's surprisingly difficult. One of the most difficult exercises for beginning pianists is to play different rhythms with each hand—part of the complexity underlying Chopin's superb miniatures. Scott Kelso, in his wide-ranging book, *Dynamic Patterns*, discusses many variations on this theme, including a test used by the Juilliard School of Music in the United States. The test is really to see how well a potential student can maintain a rhythm in the presence of interference from a metro-

40

nome. How well can the student resist the powerful drive to synchronisation? In fact music is full of synchronisations, from dancing to heavy beats to the difficulty of playing the polyrhythms of Stravinsky's *Rite of Spring*.

An even more subtle phase-locking happens among young women. Sue, Jill and Sarah share a house: they moved in at different times, had never met before, and each spends as much time with her own circle of friends as the three of them do together. Yet after a few months of living together, their menstrual cycles phase-lock. What is the coupling mechanism? One suggestion is pheromones, the sex hormones well known elsewhere in the animal kingdom. So the coupling is very slight, almost subliminal, but achieves synchronisation all the same.

As the sun goes down, as the town lights up and the revelry begins, a party of a different kind is just beginning. Away out in the wilds a firefly lights up, then another and another, until the sky is spotted with bright little dots of light. Then slowly these little pulsating dots start to phase-lock. Before long there is just one coherent flash as the fireflies turn on and off together. The coupling method is visual. So why do the fireflies synchronise like this? Could there be an advantage?—it might make it harder for a predator to focus on any one firefly. There might be an explanation of this kind, but in fact it isn't necessary. These dynamical effects occur spontaneously in oscillator systems and in fact are difficult to avoid, as we shall see now.

Animal locomotion

If you watch a cat moving around you'll see several different styles: at the slow end there is paw by paw stalking, a very slow walk. Beyond simple prowling there is a half-speed trot which finally becomes a gallop in

41

which front paws and back paws move together. Until you read a book such as Kelso's, it may never occur to you that a cat's movement might be anything other than under conscious control. Surely the cat decides whether it wants to trot or gallop. Not necessarily so, according to Kelso.

The transition between gaits can be accounted for by simple dynamical models. For instance, Kelso describes a simple experiment, finger walking. Try 'walking' on the desk with two fingers of one hand; that is, allowing index and middle finger to alternatively hit the desk. Now do both hands together, but slowly in antiphase, so that opposite fingers are down at the same time on the two hands. Easy? Now gradually speed up. At some point most people will suddenly break into the symmetrical pattern, like a cat galloping, where the fingers hit the desk together.

From ants to orcas

Ants present a source of fascination missing from many other insects. Although the destructive force of giant soldier ants is awesome, it's the complexity of their society that really grips us. How do creatures with such small brains and minimal senses achieve complex domestic arrangements and coordinated foraging? In some ways our education tends to mask the simplicity of what goes on.

An experiment with young children illustrates this. They were asked to work out how to search for gold on an alien planet, having got there by spaceship. All their solutions emphasised communication with the spaceship and central planning and direction. But ant colonies don't work like this. There is no central planning. But ants are very cooperative via signals (scents) between individuals.

Edmund Wilson is an expert on ant and ant colony behaviour. However, in 1975 he risked widespread opprobium by publishing a book entitled *Sociobiology*, in which he attempted to extend what he knew of ants and other animals to the anthropological domain. The parallels in behaviour across widely different species are amazing. In some of the most exciting and daring wildlife photography in the series *Life on Earth*, an orca, a killer whale, is seen taking a seal off the beach and then proceeding to play with it in just the way a domestic cat plays with a mouse. Same behaviour pattern, but the only thing in common between the two predatory animals is black and white markings! This is in some ways the essence of a science of complexity. What surface phenomena can we strip away to get down to the underlying forces?

Craig Reynolds' 'boids' are the simplest artificial bird imaginable. They are just simple icons on a computer screen with a few simple rules of interaction: they shouldn't get too close, should fly at roughly the same speed, and so on. Yet a flock of such boids flies in formation just like a flock of geese, the leader changes at regular intervals and they fly around obstacles just like the real flock. So here we have a phenomenon which looks the same either in a very simple simulation or in a complex organism. Is this then a scientific model? Is this the way to model biology in just the way that physics starts off with simple models? Such philosophical considerations will concern us later.

To see this philosophical issue more clearly, pseudopotential theory in solid-state physics illustrates the transition from simple to elaborated model. The first models of the properties of metals were derived from very simple quantum-mechanical models. A metal contains many atoms in a fairly regular lattice; the atomic nuclei remain anchored to the lattice points, as do most of the electrons. But some of the electrons are essentially free to roam around. Unfortunately, quantum mechanics doesn't allow any simple cancellation of electrons (negatively charged) with protons in the nucleus (positively charged). Nevertheless, the early models of solid-state physics threw away these unpleasant considerations and simply ignored almost every salient interaction in a metal! The results were excellent, but nagging doubt remained as to why this should be so successful. Deeper understanding acquired over subsequent years led to a new concept, the *pseudopotential,* which was not only quantum-mechanically rigorous but also explained in a simple way why the earlier models worked so well.

Returning to boids, we have a simple, successful model of very complex organisms. But theoretically we hit a snag immediately. In physics, measurements are frequently extremely precise, facilitating detailed comparison of theories. Biological data rarely approach this precision, making the assessment of models much more difficult.

The boid model is an algorithm, a computational procedure for determining outputs (movement) from inputs (proximity of other birds). The bird's brain achieves this same approximate mapping between these two domains. But as we discuss later, it is hard to say that the brain implements a particular computer algorithm. How we go from these simple models to an explanation in biology is still far from clear.

Are animals conscious? We all tend to assume that our pets have identities and some sort of consciousness. In fact Donald Griffin, Harvard ethologist, details extensive human-like behaviour in many animals. Beavers, for example, build dams in teams but display adaptable behaviour in emergencies. Griffin cites examples where they will *reverse* their normal practices where necessary in situations they have never seen before.

The beaver of course has a much bigger brain than any insect and some level of sophisticated communication. But now we can turn the question the other way around. How much of beaver behaviour can be explained by simple mechanisms as opposed to conscious decision making?

A current craze, which originated in Japan, is the *Tamagochi*. This is an *extremely simple* processor chip inside a little egg-like case with a computer screen. It's a computer pet. It whinges when it wants 'feeding' or needs 'attention' and has a handful of other pet-like behaviours. You need only a few attributes of cuteness, it seems, to make a machine seem to come alive.

History seems to have a cyclical character. Peace and war alternate, customs come and go, political structures wax and wane, with no still point in the turning world. Predicting the course of human society (or, even better, changing it) has so far been the stuff of novels, such as the theme of Isaac Asimov's massive *Foundation Trilogy*. In Asimov's novels the Selsdon Plan was a detailed map of social activity within the galaxy, of mathematical precision provided the numbers were large enough. Of course it didn't work because of evolutionary events unpredictable at the outset. It's not obvious that we are any further advanced.

It is disturbing, however, just how little use analysis is in controlling human events on a large scale. The 1990s war in Yugoslavia, in which the country blew itself apart, is perhaps a sad case in point. Here we have a fully developed Western society, as technologically advanced as anywhere. We have universities, television, free access to books and newspapers. Above all we have intense international scrutiny, discussion and support. All of this has achieved nothing.

Thus we might speculate that the way forward is not to try a predictive model at the macroscale level or to focus on outcomes—such as the strife in Yugoslavia. Instead we have to focus on the local interactions: change these and the rest will follow. We might say that there is pressure towards large-scale behaviour—war, peace, anarchy—which is driven from within and very difficult to control from without.

THE MORE THE MERRIER

We have seen that interactions produce complex behaviour. It doesn't really matter what the interaction is between. It could be something as complex as a person, as simple as an ant, or not a living thing at all. It is convenient just to call it an *agent*. Now we should look in more detail at how all this comes about. So for the time being we are going to leave animals and human beings and retreat into abstraction, mathematics without formulae.

We naturally tend to think of individual human behaviour and of human society as the most complex on earth. The human brain, as John Eccles, Nobel Laureate in Medicine, remarked, is the most complex thing in the universe! Yet one of the salutary lessons of the study of complexity is that it can arise from seemingly very simple beginnings. Perhaps we need a definition of complexity, but this will have to wait for a while, since a good definition is surprisingly hard to create. We need some experience of complex systems first.

Let's begin with Grid World, an imaginary civilisation on Jupiter, where the gravity is so strong that the inhabitants, Griddies, are as flat as pancakes and are literally unable to move. In fact they are rather like plants, but gravity stunts their growth. Fortunately, in Grid World the nutrients come from underground through a series of pipes spaced out at regular intervals in a nice orderly grid. Unfortunately, food is limited, and Griddies quickly die if they are surrounded by other Griddies on all sides.

The mating ritual in Grid World is pretty strange. It's a ménage à trois, but the act of reproduction is three-way with the three parents lined up in a triangle, and the offspring immediately appears in the middle, ready to tap into a food source. No suckling by the parents here. Griddies are also quite social; a Griddie left by itself soon dies. Griddipologists, the scholars of Griddie society, will tell you of great cultural diversity. Just as starships restrained by the speed of light go through generations of astronauts before reaching destinations light years away, so individual Griddies don't witness the great distances travelled by their offspring. But certain patterns of Griddies move like tribes from one place to another, preserving exactly the same shape as they move.

We won't go into any more detail about their society and customs for the moment, but instead leap forward to the seminal work of Griddie scholar Johann Konweg. The breakthrough was the realisation that, despite the depth and sophistication of the Griddie soul, the patterns of activity in Griddie society could all be modelled, on a piece of graph paper with a pencil and an eraser: a Griddie was just a square coloured in. Of course Griddie scholars use computers to look at social patterns these days, and this is exactly how these figures were drawn. We shall come back to Grid World later, because this extremely simple system displays some of the most profound ideas of complexity.

John Conway's Game of Life, a purely mathematical creation, is the underlying model for Grid World. But the message is a profound one. Complex behaviour in social systems might have very simple underlying structure. The drive to study complexity comes from the hope that there are general principles underlying many or all complex systems.

The Game of Life is very much a visual experience. Computer graphics have been the key to watching complexity unfold and we turn next to the use of computer simulation and models.

OF WORLD MODELS AND MODEL WORLDS

In 1968 a small group of scientists met in Rome to discuss the future of humankind. This was no idle talkfest. The aim was to determine just how serious were threats such as overpopulation and pollution that now hang over our future. Out of this meeting there arose an informal organisation—the Club of Rome. In 1970 the club commissioned scientists at the Massachusetts Institute of Technology to develop a simulation model of the limits and consequences of continued worldwide growth. The model incorporated modules dealing with population, agriculture, resources, industry and pollution.

The result were summarised in the 1972 book *The Limits to Growth*. The outcome was startling. Their 'standard run', which made forward projections from current trends, predicted a world catastrophe by the year 2050. Food and resources would be in short supply, pollution would peak and population would crash.

Needless to say, such dire predictions caused a sensation. Numerous critiques appeared, pointing out flaws in the model. Yet for all its faults the Limits to Growth model achieved its greatest aim. The stated goal of the modellers was:

[To] foster understanding of the varied but interdependent components—economic, political, natural and social—that make up the global system in which we all live; to bring that new understanding to the attention of policy-makers and the public worldwide; and in this way to promote new policy initiatives and action.[8]

In this sense the project succeeded. Up until that time warnings of a global environmental crisis were generally treated as fantasy. The Limits to Growth model started people thinking seriously about the future of the planet as a whole.

The Limits to Growth exercise is a good example of using a simulation model to study the behaviour of a complex system. Simulations are computer programs that recreate features of the real world. For instance, as we discuss elsewhere, the mathematics of astronomical bodies is unsolvable for as few as three bodies, such as the Earth, Moon and Sun. So the only way to model the formation of (say) a galaxy is to represent it in a simulation.

The importance of simulation models is that they allow us to perform *virtual* experiments. This ability is especially important where it is just not possible to perform the experiment concerned for real. For instance, we cannot heat up the Earth by several degrees and watch what happens. Even if we could, the risk of destroying civilisation in the process would make it a risky experiment, to say the least! However, by running a simulation *scenario* we can look at the consequences of runaway global warming without the risk of a real experiment. Aircraft manufacturers now prefer to use simulations of aircraft design before placing a design inside a wind tunnel. Likewise, trainee pilots can safely practice takeoffs and landings in a simulator before they have to face the real thing.

Previously we saw how John Conway's computer game LIFE might be adapted to model the hypothetical Griddies on Jupiter. These models are examples of *cellular automata*, or CAs for short. They are one solution to the problem of simulating processes that are spread out across space. A CA is really a grid of cells that are spread out in 1, 2 or 3 dimensions.

In two dimensions we can represent a landscape by carving up space like a chequerboard. Each cell represents an area of space. The cells are all computers (*automata*). They are identically programmed. So at any given location the state of the corresponding cell tells us something about the objects or properties to be found there.

For example, the CA grid might represent an area of the landscape and its cells contain information about such features as the local plants, elevation and soil moisture. If we program the cells appropriately then the model might simulate processes such as tree growth and seed dispersal within a forest.

Perhaps the most important feature of a CA is that the cells in the grid interact with one another. For example, if a plant dies then to replace it we can simulate seed dispersal from all the plants lying within some *neighbourhood* of the cell concerned. These neighbourhood interactions make CA models non-linear.

Probably the best known CA models are simulations of global climate change. The so-called *general circulation models* (GCM) represent the world (or particular regions) as a large grid. Each cell in the grid represents an area on the earth's surface and the atmosphere above it. To model global climate they simulate major processes such as air movement and heat and moisture transfer within and between cells. The cells are classified according to the dominant type of environment present within them (e.g. ocean, desert, pastoral, temperate forest). Critical parameters, such as albedo (the fraction of solar radiation that is reflected) and evaporation, are assigned values for each type of environment.

In the Adams novel, *The Hitchhiker's Guide to the Galaxy*, the Earth itself is portrayed as the ultimate

computer. Unfortunately real-life computers have somewhat less capacity! Even on a large, modern
supercomputer, model builders have to make many
simplifying assumptions just to reduce the problem to
one that the machine can handle. It is just not possible
to represent the entire system accurately. To understand
what this might mean in a global change model, try to
imagine a giant leaf, 100 kilometres by 100 kilometres
in area. This is literally how some simulations represent
the forests within a cell of their grid. It will come as no
surprise then to learn that the models can be extremely
sensitive to these sorts of assumptions. This is one of
the reasons why different studies of global change can
give inconsistent results.

Given the sort of unreliability we have just alluded
to, many people might throw up their hands and say
we should give up (many do!). But all is not lost. As
with the Limits to Growth model, simulations are still
useful even if they do not accurately predict what will
happen in a system.

To understand this point let's take the problem of
predicting the spread of a wildfire. On the face of it this
looks to be a much simpler problem. Here we set up a
grid in which the cells represent fuel. In different
landscapes this fuel may be grass, or bushes or trees.
To simulate a fire we set one or more cells 'burning'
and the computer works out how the fire spreads from
one cell to another.

So far so good. Unfortunately, fire-spread models
come unstuck when we try to predict what happens in
a real fire. This is not because we don't understand
fire-spread at all. It is because of two problems: inadequate data, and the sensitivity of the system. Suppose
that we want to model the spread of a fire through grass.
Then the essential problem is that we need an accurate
map of where the grass is. Now, a grassland may look

51

Figure 3.1 **Fire control scenarios in a CA model of the spread of a wildfire. The map is a digitised aerial photograph of a landscape, with roads, trees and fields. The fire-front is seen here as a black line.**

like a flat carpet from a distance, but it is composed of millions of individual grass tussocks of varying size, density and composition. Even to gather this sort of data for an experimental area is a huge job, and unfortunately wildfires never oblige by burning on experimental lots! But even if they did there is the problem of fuel moisture. The danger from a fire depends not only on how much fuel is available but also on how dry it is. Even if we were to go out and measure the moisture, we'd find it changing constantly. Also changing is the wind speed and direction, which often drives a fire-front.

Compounding this problem of poor data is the problem of sensitivity. The pattern of fire-spread is simply very sensitive to fuel load and moisture. If the grass is slightly denser and drier in one area then the fire will spread faster. The effect is even worse if the fuel is patchy, with (say) large gaps between bushes or trees.

We have described this fuel problem in some detail to make it clear that in many cases the exact pattern of fire-spread is literally unpredictable. However, this does not mean that models of fire-spread are useless. Far from it. They are simply not good for on-ground fire-fighting. Their real use is in asking what-if questions. What if there is a sudden wind change? Would a fire-break along a ridge be adequate? Would off-season burning reduce the risk for campers in the summer?

From virtual worlds to artificial life

This serious-minded 'game-playing' with cellular automata and other systems is one aspect of a booming new area of research called *artificial life*. Artificial life (or *alife* for short) aims to understand living systems better by abstracting them and simulating them in computers and other media. Besides cellular automata, alife looks at many other kinds of systems as well. Many of the ideas we shall encounter in this book really fall under the heading of alife. For instance, we can enrich a CA model by laying over it roving *agents*. These are effectively programmed automata, but instead of remaining in one spot they can move around, just as (say) a deer might wander through a forest, or a fish swim around in a coral reef.

An important method in science is to generalise from specific cases. Traditionally this has been the role of mathematics. For instance, if we can write an equation

Figure 3.2 **Examples of 1-dimensional cellular automata. Each system starts off as a row of cells (at the top of each block), with a random arrangement of black and white cells. The rows below show the new states of the cells in the CA. That is, each column follows the history of changes in the colour of a single cell. The example here represent the classes identified by Wolfram. At top left is a CA in which a brief series of transient states leads to a constant system (all cells white). At top right is an example of cyclic behaviour. The examples on the bottom row show more complex, chaotic behaviour.**

to model the growth of a population, then by studying the properties of that equation we learn not just about the population in question but also about any other system for which that equation is appropriate.

The same holds true for simulation. As we have seen, cellular automata provide models for many different systems. So from games like LIFE we can learn a lot about the behaviour of many different kinds of system.

Research on CAs really has several aims. On the one hand the models can be seen as artificial worlds. But at the same time they can be seen as models of parallel computing. As we have noted, each of the cells in a CA is a little computer. So regardless of what natural system we simulate with a CA, its grid of cells also mimics a parallel computer. It is recognition of this fact that motivates a lot of the research on CAs. A couple of decades of research on cellular automata have taught us a lot about systems of interacting cells. Stephen Wolfram, for instance, found that the behaviour of CAs fell into several general categories. All of these are visible in the computer game LIFE.

The first, and simplest, behaviour is for a system to 'freeze' into a fixed state, after which it never changes. In the game LIFE, for instance, several patterns of living cells rapidly die out, leaving behind an unchanging sea of dead cells. Others form fixed patterns that never change.

Slightly more interesting are patterns that fall into cyclic behaviour, periodically changing between several different patterns. In LIFE, a row of three live cells blinks: first it is vertical, then horizontal, then vertical again, and so on.

More complex is chaotic, aperiodic behaviour in which the system never settles down to a constant pattern. Finally there are complicated localised structures, some of which may propagate.

We can see examples of all these types of behaviour in a simple one-dimensional string of cells as it evolves through time. Because it is just a line we can lay each successive pattern that we see in the system side by side vertically, to form a two-dimensional pattern. In these pictures the vertical axis represents time. Any column in the picture shows us the history of changes in one particular cell. This means that a column of constant colour represents a cell that falls into a fixed state. A column of alternating colours indicates a cell that cycles. The other two types of behaviour are seen as larger scale patterns that *emerge* from the behaviour of the cells.

Subsequent work by researchers such as Chris Langton has shown that the different classes of behaviour tend to be associated with how complex, how 'hot', the CA model is. At first this idea of temperature for a model may seem obscure and technical. In fact it's fairly simple.

As we saw earlier, the behaviour of cells in a CA depends largely on its neighbours. One way to describe temperature is by asking whether the neighbourhood of a cell forces it to change state. If most arrangements of cell states in a neighbourhood will leave a cell unchanged then the model is very 'cool'. Cool models are usually associated with the first two of the above types of behaviour. When the model is run, the few neighbourhoods that force a change of state do so, after which the pattern of cells freezes. In a very 'hot' model, most neighbourhoods force cells to change state, so the pattern in the CA bubbles away furiously and can never settle down. So hot models lead to the third type of behaviour. The final type of behaviour seems to be associated with models that lie in between the others, in the border region between very ordered behaviour (fixed states and cycles) and chaotic behaviour.

This final type of model seems to exhibit the most interesting behaviour. That these models seem to lie in

Figure 3.3 **Phase changes in the behaviour of a CA model. The three models shown start off with the same random arrangement of black and white cells but soon produce the patterns shown here. The model on the left has low complexity and is 'cool': it changes the colour of only a few cells, so the system 'freezes' before any interesting patterns emerge in the array. In the model on the right, its behavioural complexity is high and it behaves much like a 'hot' gas: it changes the colour of almost every cell. Its behaviour is essentially chaotic. No patterns can emerge because the grid is constantly 'boiling'. The central, medium model lies at the 'edge of chaos': its behaviour is richer than the model on the left, but much 'cooler' than the model on the right, so pattern elements, such as the lines visible here, do emerge in this model.**

COOL   MEDIUM   HOT

a region between order and chaos has led to the provocative name *edge of chaos* to describe them.

This idea will occur through the remainder of the book. There are special properties of *connectivity* and *dimensionality* characterising this model. In fact, complex adaptive systems may evolve so that their behaviour lies at the edge of chaos, a startling conjecture that is one of Stuart Kauffman's many brilliant insights and can be found, with many other similarly profound theories, in his book *Origins of Order*. Kauffman is well known in alife circles. A distinguished biologist, he has moved in recent years to the study of complexity, principally at the Santa Fe Institute.

CHAPTER 4

THE IMBALANCE OF NATURE

On 5 November 1983 the *Weekend Australian* carried a startling headline: REEF PARK A FARCE IN FACE OF STARFISH THREAT.

The 'Crown of Thorns' starfish—*Acanthaster planci*—was destroying the Great Barrier Reef. The largest coral reef in the world, it seemed, was doomed.

Over the next two years the media carried many similar stories about the killer starfish. TV news showed footage of distraught tour operators desperately removing starfish by the boatload. Outraged scientists appeared on screen deploring the lack of government action. To the person in the street it seemed that we were witnessing the demise of one of the great natural wonders of the world. Millions of killer starfish were swarming over the reef eating the living corals. When the corals were gone, the fish and other animals departed. What was left was like an underwater desert, grey and lifeless.

Within the scientific community a savage controversy erupted over the nature and cause of the outbreak. One side claimed that we were witnessing the end of the Great Barrier Reef; the other side insisted that the outbreak was simply a natural phenomenon. For years the scientific debate was as fierce as the starfish themselves.

WHY IS A STARFISH LIKE AN ATOMIC BOMB?

As is usual in such polarised debates, the truth, as it eventually turned out, was not nearly so simple. With an election looming at the end of 1984 the Australian Government threw money at the problem. It commissioned AIMS, Australia's prestigious Institute for Marine Science, to carry out a survey of the problem. Roger Bradbury, the Institute's chief ecologist, led a team of more than 50 young scientists who scoured the Great Barrier Reef (GBR) looking for starfish. After several years of hard work the answers began to emerge.

The GBR, it turned out, was not dying after all—not yet, anyway. Even so, an event of enormous magnitude certainly was taking place. To understand this event we must first understand the nature of the GBR.

Most people think of the Great Barrier Reef as a gigantic wall of coral that runs all the way down the Queensland coast—nature's equivalent of the Great Wall of China. It is the obstacle that nearly wrecked Captain Cook's ship, *Endeavour*, during his voyage of discovery in 1770. Over 2000 kilometres long and millions of years old, it is the largest coral reef in the world. It is also the home for thousands of marine species.

The GBR is all of the above things and more. But one thing it definitely is not is a solid wall of coral. In reality it is a chain of reefs, thousands of them, surrounding islands, atolls or shallows that litter the coastal waters. The GBR is a series of reefs, each one separated from the others by deep water. Up until the time of the AIMS study everyone studying the problem had been mesmerised by the scale of destruction on individual reefs. Bradbury realised that, to understand what was happening, his team had to study COT (Crown of Thorns) outbreaks as a system-wide phenomenon.

60

A coral reef is a complex ecological community. It consists of hundreds of species, including fish, sponges, crabs and many other kinds of plants and animals. But ultimately all of them rely on the coral reef for their existence. The corals themselves are communities made up of millions of tiny organisms. A coral is not a single organism at all but a symbiotic relationship between a plant and an animal. The massive coral domes and branches we see are the by-products of this symbiotic relationship. All of their remarkable activity depends on sunlight and on water. The algae need light to function, so corals cannot grow in water that is more than 30 metres deep. Nor can corals survive prolonged exposure to the air. So they are confined to the shallow waters near the coast.

For bottom-dwelling corals, the deep water between islands is akin to a desert. No corals can grow in the darker waters, so no reefs form. Where coral reefs do abut deep water it is because their reef-building has kept pace with rising sea levels over thousands of years.

So when TV cameras show pictures of total destruction on the reef, what they are showing is the destruction of a single reef: a local disaster. Reefs a few kilometres away may be totally unaffected. At the time that the COT problem first came to notice, no one knew just how many reefs were affected, or how badly. Most important of all, no one knew whether the reefs would ever recover. Alarmists claimed that the reefs were doomed, but was that really the case?

The AIMS survey soon revealed some intriguing patterns. First, the destruction was patchy. Some reefs were devastated, but others were untouched. Moreover, there were no half measures: either a reef was totally destroyed or there were virtually no starfish at all. Nowhere were there large numbers of starfish living in

61

harmony with the reef. This was one case, it seemed, where the balance of nature did not exist.

What was going on? Did some reefs have special properties that made them immune? Were some reefs more susceptible to attack than others? More surveys, and accumulating historical data, began to reveal the answer. When the scientists plotted maps of outbreaks against latitude they made a startling discovery. It appeared that a wave of starfish outbreaks was moving down the GBR from north to south. This pattern soon came to be known as the 'Reichelt Wave', after Russ Reichelt, a member of Bradbury's team and later director of AIMS, who first noticed the pattern. The wave of outbreaks moved south over a period of years. The wave pattern was weak in the north, where outbreaks seemed to be more sporadic, but the pattern became progressively stronger as it moved south.

Once scientists had found a systematic pattern in the outbreaks, they began to appear more and more like a natural event. As the scientists gathered further historical data, it became apparent that the GBR had been subjected to waves of outbreaks many times in the past. It even began to appear that they might occur with a predictable frequency. Each wave of outbreaks would run its course in a few years, after which the GBR would gradually return to its usual healthy state.

Having identified the pattern, the scientists next turned their attention to the process. Several crucial questions still remained unanswered. In particular, how did the outbreaks spread? Did armies of starfish march across the ocean floor from one reef to the next? Tracking individual starfish is notoriously difficult. Unlike fish or birds, you cannot tag a starfish. Try it, and the starfish simply ejects the metal from its body.

Despite these difficulties, the biologists soon had their answer. During the summer spawning period the starfish

62

released their young. The waters filled with clouds of minute larvae. They spread in such vast quantities that even the hungriest predators could eat their fill and yet have no noticeable effect on the numbers. Within hours the clouds drifted away from the reefs where they formed and moved out into deeper waters. Currents then picked up the larvae, driving them away from their source. Oceanographers have long known that currents wash the entire length of the GBR from north to south. These currents now carried the larvae south. Within days millions of them had settled. Another reef was doomed. The cause of the Reichelt Wave was now clear: outbreaks on any given reef eventually lead to a larval cloud, which then seeds outbreaks on other reefs, and so on.

The only remaining question was how the outbreaks started in the first place. Obviously the process has to be seeded by outbreaks in the waters north of Cape York, but how? Without massive influxes of starfish larvae, why should an outbreak suddenly start? Is it chance? Or is it something more serious? The most contentious theory is that overfishing reduces the predators that normally feed on starfish larvae and young starfish. The issue is hotly debated to this day.

At least now we know how the outbreaks spread. At the time of writing, fresh outbreaks are again being reported in northern reefs. If the above theory is correct, then scientists should be able to observe another wave of outbreaks over the next few years. Most importantly, the process is now predictable. We know what to expect. We need not be quite so alarmed the next time starfish begin appearing in a prime tourist spot. Perhaps we can even prevent it.

As it now appears, the COT outbreaks are what scientists refer to as an 'epidemic process'. When one reef succumbs it is akin to being infected with a disease.

Moreover, it infects other reefs in the same way that colds and flu spread through a schoolground or an office.

All through history, humans have been subjected to plagues of one kind or another. The AIDS epidemic is but the latest. During the Middle Ages, Europe was ravaged time and again by plagues that swept across the continent, usually starting from somewhere in the Middle East. Towns were worst affected. All it needed was for a single infected individual to arrive in town. By the time the unsuspecting inhabitants knew what was happening it was too late. Within a few weeks trundle carts would be plying the streets collecting the dead by the hundreds. Many rural villages, knowing their peril, would close themselves off from the world and take the extreme measure of killing any strangers unlucky enough to arrive seeking shelter for the night.

So the COT outbreaks are a typical epidemic process. Many other processes act in this way too. For instance, we tend to think of fire burning through a landscape as almost a living, growing thing. Yet fire too is an epidemic process. Fire spreads from one tree, one bush or tussock of grass to the next. If the bushes are too far apart, then the fire cannot spread. Though less obviously, water seeping through soil, and weeds spreading from paddock to paddock, are also epidemics of a kind.

But perhaps the best known, and indeed most infamous, epidemic is the nuclear chain reaction that sets off an atomic bomb. An atomic explosion occurs when a neutron collides with a uranium or plutonium atom, causing the atom to split in two. In the process the atom fires out two new neutrons. Each of these goes on to hit other atoms, which also split and send out still more neutrons. The explosion is caused by the resulting heat, which is released at a rapidly increasing rate as the number of splitting atoms doubles, quadruples and so on. The process spreads in essentially the same fashion as any epidemic, including the spread of COT starfish.

And that is why a starfish, or at least an outbreak of them, is like an atomic bomb. So what?, you may ask. How does it help us to know that a starfish is like an atomic explosion? Surely the two things are even more different from one another than chalk and cheese?

In truth it does matter that starfish outbreaks, plagues, fires and atomic fission are all related processes. For in understanding this similarity we can learn more about each one than we ever could by studying it on its own. For instance, before they could build an atomic bomb the nuclear physicists at work on the Manhattan Project first had to learn about the properties of chain reactions. Some of these discoveries apply directly to other epidemics. For instance, to work at all an atomic bomb has to have a certain 'critical mass' of fuel. If the lump of uranium is too small, then nothing happens. The alpha particles emitted from a dividing nucleus simply shoot off into empty space, missing all the other atoms completely. The bomb makers exploited this property by dividing the nuclear fuel in two. Each piece by itself was too small to explode, but when slammed together by a conventional chemical explosion the two formed a critical mass.

For similar reasons the classic approach to infections is isolation. Isolate the cases and you avoid forming a 'critical mass' of human 'fuel'. The point is that, in identifying this common property, one system can offer insights about another. For instance, perhaps some critical number of COT outbreaks in northern waters is necessary before a wave of outbreaks can get under way. If so, then one way to avoid starfish outbreaks is to reduce the incidence of those initial outbreaks. Curtailing fishing may be one answer. Short of that, however, applying preventative containment to seed outbreaks is likely to be far more cost-effective than a Bandaid approach of trying to rescue dozens of high-profile reefs once the process is under way.

Phenomena, such as epidemics, that characterise complex systems permeate the whole of the natural world. It is possible that they even include mental phenomena. Are flashes of insight and bright ideas examples of neural epidemics? Such questions illustrate the extreme diversity of complex systems. Yet, given this diversity, there seem to be underlying patterns which beg for some global theory or paradigm. An almost universal characteristic of a complex system is a set of agents of some kind which interact in some way. These agents may themselves be complex, but the properties of the system are not easily predicted from a knowledge of the agents and the way they interact. Only step-by-step simulation of the system will reveal its behaviour.

One very important parameter of complex systems is connectivity: which agents talk to which others. What is interesting, and extremely powerful as a theoretical tool, is the discovery that it is just the existence or lack of connections that is significant, not their strength or dynamics. We look at the implications of this more closely in Chapter 6.

COMPLEXITY IN THE LIVING WORLD

One of the greatest challenges facing humanity today is to manage and conserve the world's living resources. Enlightened management has to be based on accurate, up-to-date information. The first problem is simply to learn what plants and animals are out there.

How many species are there? Science is not able to answer this simple question with any certainty. Estimates by different methods put the world's biodiversity at anywhere from two million up to 100 million species. A best guess is probably in the vicinity of ten million. The main reason why we are so vague about how many species there are is that only a tiny proportion of the world's plants and animals have ever been described scientifically. One UN estimate is that the task will take at least another 600 years to complete. The fear is that in rainforests and other ecosystems, whole communities of species are being wiped out before we even know they exist.

One way to estimate species numbers is to scale upwards from what we know. When scientists survey a new area for the first time, they are likely to find many new species. Later surveys inevitably find fewer and fewer new species as the stock of new species dwindles. This is a case of diminishing returns: it takes a lot of effort to find the last few species. On the other hand the rate at which new species are found falls off in a predictable way. So if we compare effort with the rate of discovery, then it is possible to predict how many species are still unknown. This kind of prediction indicates that there are relatively few unknown species in some well-studied parts of the world (e.g. most of Europe), but in other places there are still areas largely untouched. Even in developed countries, there are many blanks on maps of regions surveyed for biodiversity. In Australia, for instance, scientists have actually produced maps to show where specimens were obtained. For central Australia, these points formed a perfect map of the roads through the Australian outback! The rest of the continent is largely unsampled.

Many national and international projects are under way to compile lists of species for different parts of the world. One source of data is to catalogue specimens in taxonomic collections. This may not seem like an onerous task. However, to understand just how enormous a challenge we face in databasing biodiversity it is worth looking more closely at some examples, so we can know what is involved.

The Australian National Insect Collection is housed in a large building in Canberra. The main collection resides in row after row of cabinets that are spread over several floors. Each cabinet contains many drawers, and each drawer holds dozens of specimens. In all there are over eleven million specimens in the collection. Electronic cataloguing of this vast array of specimens began in the early 1990s with a single technician entering data off the specimen labels. It was estimated that the technician would be able to capture data for about 50 000 specimens in a single year. This effort sounds impressive until we consider the size of the collection. At that rate it would take over 200 years to record all the specimens!

Added to the sheer problem of recording the mountains of data already available is the problem of making sure that the data are reliable. Errors can creep in at any stage of the recording process. The original collector could make a mistake about where the specimen came from. The specimen could be misidentified. Likewise the scientific name could have changed and not been recorded. Then there are typographical errors, either in writing the original label or in entering it on a computer. The most useful data about location are the latitude and longitude. Early scientists were not aware of the needs of modern science. So crucial data such as location are sometimes missing. More often they are given in some off-the-cuff manner, such as 'five miles north of Gundagai'.

A little anecdote might help to drive home how serious all these errors can be. One of the first data sets to be compiled from specimen records, as described here, were lists of endangered species for Australia. When these data were first plotted as points on a map, it was found that several hundred specimens had been mislocated in the ocean! If this sounds bad, remember that these records were relatively easy to fix. Even more worrying than points mislocated in the sea are points that are wrong but not *obviously* out of place.

WHAT IS ECOLOGY?

Ecology is the branch of biology that looks at communities of plants and animals in the environment. Order in ecosystems arises from two different sources: the external environment and internal interactions. Ecology has different ways of looking at these two aspects of ecosystems. One way, *autecology*, looks at each species and how it interacts with the environment. What is its distribution? What sort of climate does it prefer? How does it reproduce and spread? Besides looking at interactions between organisms and their environment we also need to consider relationships between different kinds of organisms. *Synecology*, the study of living communities, looks at the ways in which populations interact.

The environment shapes natural communities in many ways. Plants cannot grow where the temperature is too extreme; animals cannot live where there is no water. There are endless variations on this theme. Because most environmental properties vary from one place to another, landscapes are a central concern in ecology. For instance, the temperature grows colder the higher one goes up a mountainside. Tropical heat at the foot of a mountain can give way to freezing cold at the top. This change gives rise to effects such as the alpine tree-line. In similar fashion, rainwater quickly runs off hilltops and into gullies where it becomes concentrated in creeks and rivers. This is why rainforest tends to be concentrated in valleys, whereas trees that prefer dry soils, such as pines and eucalypts, congregate on upper slopes and ridges.

Not only do properties of the landscape affect what can live where; processes that go on in the landscape have a big effect too. Floods, droughts, storms and fires all limit the kinds of plants and animals that can live in any given area. Fire in particular is an extremely powerful force for shaping communities in a landscape. In many parts of the world, native peoples deliberately light fires as a tool to manage the landscape for their own ends. One effect is to clear the underbrush, making hunting much easier. Another is to promote the fresh growth of grass and edible plants as food for game species.

The other major force that shapes ecosystems consists of the interactions between organisms. These interactions are often less obvious than environmental factors but in the long run have just as great an influence over what lives where. When we think of interactions between different living things the images that spring to mind tend to be lions chasing zebras, cattle eating grass or even viruses infecting human beings. As we shall see, these kinds of interactions certainly are important. But interactions between plants and animals of the same species are equally significant.

Cooperation and competition

Within any population, individual plants and animals interact in many different ways. The lyrebird dances to attract a mate. Male sea-lions fight each other for control of territory and females. Monkeys suckle and rear their offspring. Wolves cooperate to hunt in packs. Let's look briefly at some of these interactions more closely. We are all so familiar with the idea of animals raising their young that we tend to assume that parental care is a universal feature of the natural world. But this is not so. Most frogs, for example, lay their eggs and then forget them. Baby tadpoles are left to their own devices.

Parental care is really just one of many strategies that organisms can adopt towards reproduction. Ecologists have found that, in the struggle to survive in the environment, reproductive strategies lie between two extremes. One extreme strategy is to produce lots and lots of offspring, spread them far and wide, and hope that some will survive to adulthood. Locusts, starfish and other plague species adopt this strategy. So do many plant species. They rely on the odds that at least some offspring will survive. A very different strategy is to produce few offspring but to devote lots of time and effort to ensuring that they survive. Parental care—as we see in humans and other social animals—is one strategy of this kind. Other examples include cuckoos placing their offspring in the nests of other birds, and wasps laying their eggs in the bodies of living insects. Likewise fruit trees produce their seeds inside fleshy containers, which serve both to attract animals that disperse the seeds and to provide nutrients for the seeds when they germinate.

Another familiar kind of interaction between organisms is competition. Living things compete against one another for food, living space and mates. This competition provides the driving force for natural selection. More individuals are born than can be supported by the available resources. The competition for limited resources weeds out individuals that are poorly adapted and favours individuals that are *fittest* (best adapted). The crucial point is that the fittest individuals in a population tend to be the ones that survive to maturity and that then reproduce and pass on their genes to the next generation. So the best genes tend to be passed on from one generation to the next. This is natural selection.

Interactions between species also play a major role in shaping the complexity of nature. Just as competition occurs within a species, so too does competition between similar species. What are the effects of this competition?

One of the early ideas put forward about this was the *competitive exclusion hypothesis*. This theory proposed that whenever two populations compete with one another, one population will drive the other to local extinction. That is, one species disappears from the local area.

The above theory agrees with many field observations and correctly predicts what happens in laboratory experiments when (say) different microbes are left together in a flask. However, so many exceptions have been found that the idea has become largely discredited. Why have lions not driven hyenas to extinction? How are so many similar species of trees able to coexist? The answer seems to be that competition is not always as intense as it inevitably is in a closed laboratory flask. Also 'competing' populations are not always really competing. In the real world populations are spread across a landscape. One way in which similar species coexist is by dominating different territory. When this happens the only real tension is at the borders between territories.

In a sense all plants compete with all other plants. They are all vying for space in which to grow. Studies of plant distributions provide some insights into the way competitive exclusion translates into spatial patterns. One of the key workers in this area, Chris Pielou, carried out extensive studies in Canada into the ways in which plants arrange themselves spatially. One of the most interesting cases is what happens on an *environmental gradient*. This is an area where some physical property changes across the landscape. Examples include temperature declining up the side of a mountain, increasing salinity in a tidal swamp as you move closer to the sea, or decreasing moisture in the soil as you move from a gully up a slope to a ridge. Now, the thing is that each species has a particular range of moisture regime that it prefers. There are absolute limits outside which the plant simply cannot grow at all. To take one extreme, most plants die if submerged in water; otherwise our lakes and streams would be choked with plant life. At the other extreme, plants also die if left in desert soil that is completely dry. They prefer some intermediate range of moisture and some plants require more than others.

At first glance we would expect that, on a hillside (say), each plant species would occupy the full range that it can tolerate. However, this does not normally happen. What Pielou and others found was that species ranges tend to be cut short. Rather than finding (say) fewer and fewer rainforest plants as we move uphill, their distribution finishes, sometimes abruptly, because they cannot compete effectively at the ends of their potential range. This process often gives rise to very distinct zones, such as intertidal rock pools or the bands of different vegetation that we see on the side of a mountain.

So what? What good does it do to know that competition leads to sharp boundaries? There are several immediate implications. One concerns our ability to predict the distributions of species. Earlier we looked at the compilation of biodiversity data. For conservation purposes it is not enough to know that a species is out there somewhere. We need to know where. We need to be able to pinpoint its distribution exactly. For instance, if a new road is going to be built planners need to know whether it will have any impact on rare or endangered species. If we know the climate 'preferred' by a given species then we can predict its potential distribution by looking on a map for all sites in the landscape that fall within that envelope. By 'climatic envelope' we mean the range of values that the species tolerates for rainfall, temperature and so on. So if we know what areas may be home to rare species we can go and check to see whether they are found there. We can also use this approach to check for other areas that might be ideal for species that are under threat elsewhere.

Now, the problem is that competition ruins the simple approach outlined above. Because of competition, the plants and animals that we are concerned about may be completely absent from an area where they could thrive. Competition may also prevent them from becoming established in new habitats. Conversely, if we introduce a new species into an area where it competes with native species there is a risk that it will wipe out the native species altogether.

Is the cane toad an evil force?

Early in the 1930s the Australian sugar crop was under threat. The cane beetle was causing great damage in the sugar cane fields of North Queensland. In an attempt to control this curse the government introduced the cane toad from Hawaii.

Rather than the instant cure that was intended, this action turned into an environmental disaster. Not only did the cane toad eat the cane beetles and their grubs, as intended, it also ate anything else that would fit in its mouth. What is worse, the toad's skin exudes poison, which kills any predator that eats it. The cane toad quickly became a greater pest than the pests it was brought in to control. Today it occupies the entire coastal plain of Queensland as well as northern New South Wales and part of the Northern Territory. There is a very real danger that it might lead to the extinction of a host of native carnivores and other small animals.

Why are introduced species always so damaging to the environment? The fact is that they are not. Most introduced species have to struggle for survival. They cannot compete against the locals. Ask any gardener! The introduced plants and animals that we see as pests are just those species that are highly competitive. Blackberries are not a pest in Europe where they are constrained by competition with many other plants, but in Australia they are a serious weed. Conversely, the eucalypts are severely inhibited within Australia, their homeland, by poor soils and insect attack. In North America, where these restraints do not exist, some eucalypt species grow at incredible rates and are a serious weed.

Interactions between species influence the structure and function of entire ecosystems. An important problem for ecology is to study how ecosystems are 'organised'. One way of looking at this question is to ask how energy and nutrients move through an ecosystem. An important pattern is the *carbon cycle*. Plants draw carbon from the soil and fix it in their tissues. Herbivores eat the plants and absorb the carbon into their bodies. The carbon passes on to predators when they eat the herbivores. When these animals die in their turn, decomposers break down their bodies and return the carbon to the soil, so completing the cycle.

Biologists have spent a lot of time trying to model the ways in which plant and animal populations change through time. They quickly realised that, although foxes and rabbits and spiders and flies may be vastly different animals, the ways in which they interact are fundamentally identical. In each case one animal is a predator and the other its prey. Predation—one species eating another—is just one of many types of interactions that can occur between different animals. There are hosts and parasites, in which the parasite does not kill the host but, like ivy clinging to a tree trunk, benefits from the host in some way.

Cycles and feedback

In general, if there are more species in an ecosystem then there will be more interactions, and more complex interactions. The American ecologist Richard Levins

75

delights in telling stories about the dangers of interfering with rainforests and other complex ecosystems. Removing one innocuous plant could set off a complex chain of events with disastrous consequences. He argued that it was virtually impossible to model a rainforest accurately. There are simply too many variables and too many interactions to measure. Instead, he suggested a very simple, yet effective, way to study processes in these complex communities.

Levins's *loop analysis* looked at the patterns of interactions between different species. We can simplify the many kinds of possible interactions between species by asking whether they are positive or negative for each of the species concerned. A positive interaction is one that tends to help the population increase in size. A negative interaction is one that decreases the size of the population. A host species has a positive effect on a parasite. The more hosts the more parasites that can live. A predator has a negative effect on its prey—the more predators there are the faster the prey animals will die.

The power of this way of looking at nature becomes apparent when we consider whole chains of ·interactions. Take foxes and rabbits. The FOX–RABBIT interaction is negative: more foxes kill rabbits faster and lead to fewer rabbits. The RABBIT–FOX interaction is positive: more rabbits mean food for more foxes, so the fox population can increase in size. This combination of effects leads to a feedback loop. It is a loop because we can trace a series of interactions from rabbits to foxes and back to rabbits again. It is negative because the sequence contains an odd number of negative interactions.

So what does this mean? What happens if we put the foxes and rabbits together? Well, as the foxes start killing rabbits, so the rabbit population falls. But fewer rabbits means that there is less food for the foxes, so in time the fox population starts to fall. This decrease in fox numbers means that more rabbits can survive, so the rabbit population increases. But that means that fox numbers can increase again, and so on. So over time the whole system cycles, with the numbers of foxes and rabbits oscillating up and down.

This kind of system is known as negative feedback. It is called feedback because the effects of any change in one of the components eventually 'feeds back' to affect that component itself. In negative feedback, the system tends to cancel out any change in the components. So any increase in the rabbit population produces effects (namely more foxes) that tends to decrease rabbit numbers again.

Feedback is a well-known process. Engineers have studied and exploited feedback for many years. A speed limiter on a car, for instance, is a negative feedback device that is designed to counter any tendency of the vehicle to go faster than a set speed. The other type of feedback loop is positive feedback. In positive feedback the net effect of the loop is positive. This means that any change tends to be self-perpetuating. That is, the change keeps on going, and usually accelerates.

In general, negative feedback promotes stability in a system—it tends to negate change. A host of man-made devices, from speed delimiters to temperature regulators, exploit this principle. Positive feedback, on the other hand, promotes instability. It feeds on any change introduced into the system, and leads to further change. Now, in any ecosystem there will be lots of interactions between the different populations that make up the system. The question is whether those interactions will produce positive or negative feedback.

Positive feedback in an ecosystem is disastrous. Suppose that positive feedback starts affecting some

Figure 4.1 Coupled feedback loops from the Limits to Growth simulation model of world population (see Chapter 3). The arrows indicate the kind of effect (positive or negative) that different variables have on each other. We can determine the nature of each loop by counting the number of plus and minus signs within it. An odd number of minus signs shows that the loop is negative; otherwise it is a positive loop. In a negative feedback loop any increase in a variable will 'feed back' to it and tend to decrease it again.

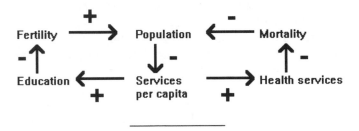

animal population. Then each change in the population will lead to further change in the same direction. If the population is falling then it will keep falling—at an ever-increasing rate—until the population crashes to zero. This is what stability theory predicts for competition. If two species (foxes and weasels, say) are competing then the effect of each population on the other is negative. Although it is not at first apparent, this situation produces positive feedback. More foxes means fewer weasels, and fewer weasels in turn means more foxes. So the system tends to magnify any change. So such a system is inherently unbalanced. Any increase in one population will perpetuate further increases, and simultaneously drive the competitor to extinction.

Now let's look at what actually happens in a large community containing many plant and animal species. Naturally there will be many interactions, of many different kinds, between the populations involved. These interactions will produce many feedback loops. Suppose that we could select the species at random. Then we would expect that about half of the interactions would be negative and about half would be positive. So the resulting system should contain both negative and positive feedback loops. Those positive feedback loops have the potential to disrupt the entire system.

In the early 1970s the Australian physicist Robert May caused a sensation in the world of ecology. In the course of many years of observing natural communities, ecologists had come up with various generalisations. One of the best known of these rules was that complex ecosystems tend to be more stable than simple ones. This principle was based on the observation that complex communities, like rainforests or coral reefs, tend to recover faster than simple ecosystems. The thought was that complexity somehow enables a system to resist change. May caused a shock by showing that in artificial systems the exact opposite is true. The more complex a system is the more likely it is to be unstable.

We can appreciate the truth of May's argument at once by thinking about feedback loops. As we saw above, systems with lots of species and lots of interactions are likely to be destabilised by positive feedback loops. Experiments confirm this intuitive conclusion. Now, it is simply not possible to carry out experiments on this matter in the real world. They would be too slow, too costly, too difficult and too disruptive. Instead we can carry out experiments on a computer. To do this we make up model communities and have the computer assign random interactions between the different species. We can then run the model to see what happens to the mixture of species that we have formed. Not surprisingly, most of the models contain positive feedback loops and collapse. That is, one or more species goes extinct in the model world that we have created. Only a tiny fraction of the model communities retain all of their species.

Experiments like these provide important insights about the nature of ecosystems. Perhaps the sort of collapse that we see in the models goes on in nature all the time. Perhaps the ecosystems that we see around us are just the combinations that do not collapse. Or perhaps systems are collapsing all the time, but we are just not aware of it. In any case, ever since May raised the question about stability and complexity, ecologists have been trying to understand whether complexity is associated with stability in ecosystems, and if so why.

The issue is not at all simple. For a start we have to ask many questions about the model ecosystems. For instance, we saw earlier that competitive exclusion does not always occur in landscapes because the competing populations are spread over the landscape and may not always be interacting directly. Other factors come into play as well. It is one thing to say that a certain set of species forms a positive feedback loop, but how strong are the interactions compared to other processes that are going on at the same time? They may be very weak. The above effects are especially important when one population is involved in several different loops.

Time delays have an enormous bearing as well. Sure, more rabbits may make life easier for the fox population, but it takes time for the foxes to breed and for their offspring to reach an age where they can start hunting rabbits. Delays of this kind tend to produce oscillations of the kind we saw before.

Despite the uncertainties that remain about the role of feedback loops in ecosystems, they do provide several strong messages. One is to emphasise just how dangerous it is to introduce new species into an environment. As with the cane toad, the history of conservation is full of stories about well-intended introductions of exotic species leading to environmental disasters.

IS THERE A BALANCE OF NATURE?

Popular portrayals of ecology always emphasise the idea of balance. They are full of expressions such as 'the natural cycle', or 'upset the balance of nature'. The idea of equilibrium—balance—is almost dogma in some sections of the environmental movement. And yet is it really true? Is there a balance of nature? Does such a thing really exist? The above discussion has already raised serious questions about the real stability of ecosystems. Ecologists essentially abandoned assumptions about equilibrium years ago. To some extent our perception of stability and balance in the natural world is an illusion. It is worth looking briefly at why this is so.

In the south-east of the Australian state of Victoria lies a vast wilderness of hills covered in forest. When one of the authors was slightly younger he used to spend a lot of time bushwalking and camping in these forests. Go out there today and the region still looks just as it did all those years ago. There are tall trees and mossy rocks and rainforest gullies clustered by the creek beds. And yet there are signs that something has changed. Poking up above the canopy in many places you can see the skeletons of enormous trees. Those skeletons were there when the author was young. They are still there today. Taken at face value, one could almost believe that they had always been there. But the truth is that they are the remains of trees killed in an enormous bushfire that ravaged the region in the year 1939. The size of these skeletons sticking above the canopy shows that the forest we see today has still not reached maturity. Although it looks stable and constant, certainly throughout one person's lifetime, those

81

stumps are a vivid reminder that great changes have taken place in the past.

The pace of change in ecosystems varies enormously. Populations of fruit flies can turn over in a few days. On the other hand we can find in California and elsewhere groves of trees that were growing before Christ was born. Events in some ecosystems, especially forests, can take so long that they look as though no change is happening at all. In short, they *look* stable. It is the time scale that deceives us.

CHAPTER 5

THE ABC OF NATURE

Philosophy is written in this grand book—I mean the universe—which stands continually open to our gaze, but it cannot be understood unless one first learns to comprehend the language and interpret the characters in which it is written. It is written in the language of mathematics . . .

—Galileo in *Il Saggiatore*, 1623

. . . like a child, we can spell out the alphabet without understanding more than a few words on a page.

—Nobel Prize winner William Gilbert,
on the genetic code.

Every age has its own way of viewing the world. During the industrial revolution the world was seen as a great machine. This view coloured people's thinking about all sorts of matters. The heart is a pump; the arms and legs are bio-mechanical levers. The solar system operates like clockwork.

Today we are in the midst of an information revolution. Telephones, TVs and computers are everywhere. They dominate the lives of many of us. Instead of machines it is now computers that colour our view of the world. The computer seems to be a model for understanding many processes in the natural world. For convenience we'll call this view the 'nature as computing' paradigm. Perhaps the most seminal image is that of the genetic code as life's program. In a literal sense there is a striking resemblance between the sequence of bases in a strand of DNA and a computer program written on a magnetic tape. But the similarity goes much deeper still. Like a magnetic tape, a DNA strand can be copied to produce new DNA, or to produce RNA which is similar but has only a single strand. The body uses RNA as a template for manufacturing proteins. The sequence of bases along the DNA chain forms the genetic code. DNA has four bases: Adenine (A), Cytosine (C), Guanine (G) and Thymine (T). RNA substitutes Uracil for Thymine. Because DNA consists of paired chains, it always contains two sequences that are complementary to one another. So, for example, the sequence AATGCTG is complementary to the sequence TTACGAC.

Two crucial processes occur when the DNA helix unwinds and tears apart. One is that each half of the chain can form the template for a complete new chain. The other is that the chain can be read to provide the data for growth and other processes. This happens when the code is copied to form RNA, which is another type of chain. RNA is a single strand copy of the DNA sequence. However, instead of Thymine it contains Uracil as the fourth base.

The crucial process is the interaction of mRNA with ribosomes to manufacture proteins. Ribosomes are almost literally small molecular machines. They attach themselves to mRNA and move along the chain. Each site they attach to serves as a template for forming amino acids. As they come off the production line, the amino acids string together to form proteins. In this process the ribosome acts very much like a tiny Turing Machine

84

(see Chapter 2), reading in the RNA like a data tape and outputting sequences of amino acids to form proteins. Several aspects of the above process are worth noting. First, as a template for amino acid production, the ribosomes use triplets of bases (called *codons*). So the codons are triplets, such as AAT or GCA. Each codon corresponds to a single amino acid. Now there are four bases at each of three positions within a codon, so there are 64 (=4^3) possible codons. However, there are only 20 amino acids, so there is a lot of redundancy. In most cases the third base in a codon is irrelevant; the first two bases determine what the resulting amino acid will be. This has led to speculation that the present code is derived from an earlier form in which codons consisted of pairs of bases.

Another computer-like feature of protein production is that, just like a computer program, there is a start codon, which tells the ribosome where to start making amino acids, and codons that tell it where to stop. Notice that because codons consist of three bases each, there are three possible places where codons can start. Also there are two possible directions. So the same piece of code can be read in six possible ways, called *reading frames*. An *open reading frame* is a segment of code between a start and a stop codon. A gene is a segment of DNA that codes for a particular function. Usually this function is to produce a protein. However, genes can also have several other roles, such as controlling the operation of other genes. An interesting side effect of having several different reading frames is that the code for different genes can overlap. Such overlaps are possible because the redundancy of the code allows some freedom for bases (especially the third position) to change (silent mutations) without altering the resulting amino acid sequence. This property of overlap is most

85

frequent in organisms, such as viruses, that have relatively little DNA.

Another strange feature of DNA is the enormous amount of *junk DNA*. Only a fraction of the DNA strand actually contains genetic code. The rest (as much as 80 per cent) contains base sequences that apparently are never read. This fact posed a great puzzle for molecular biologists. Why should organisms waste so much 'effort' on building junk DNA? Remember that DNA is present inside every cell in the body. So the amount of food and energy tied up in the production of junk DNA is considerable. It's just not like living things to be as inefficient as these figures suggest. There must be some advantage in having so much spare DNA lying around.

Perhaps the most likely answer is that junk DNA is needed to enable the reading of gene sequences. Although its primary structure is a helix, the DNA molecule also tends to twist itself up into complex knots—the so-called secondary and tertiary structure. So most of the sequence may be inaccessible for most of the time. The coils of DNA need to unravel at least slightly for copying to take place. The junk DNA may help to ensure that all of the coding regions are accessible during this process.

Another suggestion is that the junk DNA provides space for writing new genes. Overlapping genes do occur (especially in viruses). But although overlap represents efficient storage, it also inhibits variation. Mutations in one sequence would be likely to affect an overlapping sequence as well. Perhaps both of these explanations (not to mention others) are at least partly valid.

VIRTUAL PLANTS

Computer-like processes in nature are not restricted to protein synthesis and development. In particular, the idea of language in nature goes well beyond the coding of our genes. Biologists have turned increasingly to language as a means of describing the ways in which biological systems are organised.

When we say 'language' here we don't mean every-day speech. The problem is that normal language is altogether too flexible. Instead, we use artificial languages that are built specially to represent particular systems. These *formal languages* are just rules for manipulating symbols. Traditional mathematics, with all its symbols and equations, is really a formal language.

To appreciate the nature of formal languages, and the way they differ from normal languages, we need to go back to grammar school briefly:

<sentence>	→	<subject> <verb> <predicate>
<subject>	→	<article> <noun>
<predicate>	→	<object>
<predicate>	→	<adverb phrase>
<adverb phrase>	→	<preposition> <article> <noun>
<preposition>	→	on
<object>	→	<article> <noun>
<article>	→	the
<article>	→	a
<noun>	→	cat
<noun>	→	rat
<noun>	→	cheese
<noun>	→	mat
<verb>	→	sat
<verb>	→	ate
<verb>	→	chased

The above rules (a *grammar*) define a formal language. This language is simply the set of all sentences that can be derived by applying the above rules. Notice that the rules contain two sorts of items: terms in brackets (*grammatical variables*) and terms without brackets (*grammatical constants*). The derivation process consists of

87

starting from the grammatical variable <sentence> and using the rules to replace each variable until there are none left. Here is how we would derive a familiar sentence:

```
<sentence>
<subject>              <verb> <predicate>
<subject>         sat   <predicate>
<article>  <noun> sat   <predicate>
<article>  <noun> sat   <adverb phrase>
the        <noun> sat   <adverb phrase>
the        cat    sat   <adverb phrase>
the        cat    sat   <preposition>   <article> <noun>
the        cat    sat   on              <article> <noun>
the        cat    sat   on              the       <noun>
the        cat    sat   on              the       mat
```

The richness of the language arises from the number of choices that we have for replacing each variable as we go. Here are some other sentences that we can derive from the above grammar. See whether you can work out how they are derived:

the cat chased a rat
a rat ate the cheese
a cheese sat on the mat

Do not be fooled. This language is *not* English! It may look like English. In fact it is a subset of English. But there are many sentences we can make up within this language that make little or no sense at all. For example:

a mat chased the mat
the cheese sat the rat

Also the language can be very restrictive. We can make up many perfectly sensible English sentences, using the words given, that are not part of the above language. For example:

the rat ate cheese

L-systems

Have you ever looked closely at a plant? Next time you're in the garden look at the pattern of branches, leaves and flowers on a bush or a small tree. The arrangement is usually very regular (at least before damage by insects, storms and wilting). L-systems allow us to capture that regularity in a computer model.

Biologists have always been struck by the symmetry and regularity of plant growth. In the late 1960s the biologist Aristid Lindenmayer realised that formal languages could capture these patterns. Borrowing the first letter of his name, the approach that he pioneered is now called L-systems.

Figure 5.1 An L-system model of a branch. This picture was generated using the freeware program FRACTINT.

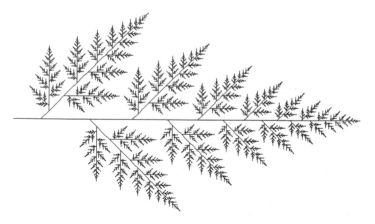

L-systems apply the above grammatical approach in order to model growth. Instead of terms about cats, rats and prepositions, they use terms that refer to (say) parts of a plant.

To understand this, let's take a simple example, defined by the following rules:

Variables :   A, B
Constants :   none
Start :   A
Rules :   A → B
    :   B → AB

This L-system produces the following sequence of strings:

Stage 0: A
Stage 1: B
Stage 2: AB
Stage 3: BAB
Stage 4: ABBAB
Stage 5: BABABBAB
Stage 6: ABBABBABABBAB
Stage 7: BABABBABABBABBABABBAB

Counting the length of each string yields the famous Fibonacci sequence: 1 1 2 3 5 8 13 21 34 . . . . This simple sequence also demonstrates the way in which iterative properties of L-systems lead to self-similarity and hence fractal growth patterns.

Growth patterns for two tree species

Simplified models are given beneath the diagrams. Names in angle brackets denote growth structures and arrows denote growth processes. The states of particular struc-

Figure 5.2 Branching patterns in *Eucalyptus alpina* **and** *Hakea salicifolia*. **The rules given in the text describe the organisation.**

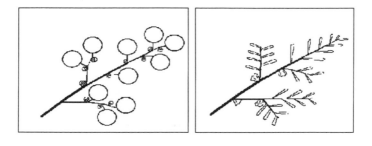

tures are indicated by colons. The models differ syntactically in the timing of fruit production and semantically in the shapes of leaves and fruit.

(a) *Eucalyptus alpina*:

```
<bud: leaf, angle>         →  <meristem> <bud: stem, angle>
<bud: stem, angle>         →  <stem> <bud: leaf, angle + increment>
<meristem>                 →  <leaf> <fruit> <bud: juvenile, 0>
<bud: juvenile, angle>     →  <bud: stem, angle> /delay=1 year
```

(b) *Hakea salicifolia*:

```
<bud: leaf, angle>         →  <meristem> <bud: stem, angle>
<bud: stem, angle>         →  <stem> <bud: leaf, angle + increment>
<meristem>                 →  <leaf> <fruit> <bud: juvenile, 0>
<bud: juvenile, angle>     →  <bud: stem, angle> /delay=1 year
```

VIRTUAL ANTS

Another popular image in computing is that of the robot. Mention the word robot and most people immediately think of a mechanical man. However, biologists are more and more aware that the behaviour of many organisms is like that of a robot. Insects, for instance, with their simple repertoire of stimulus–response actions, behave as though they are programmed to respond in a fixed way. One of the really profound differences that computers have made is that they allow us to go beyond numbers and calculations. Computers originated as calculating machines. Charles Babbage designed his 'Analytic Engine' as a tool for generating mathematical tables. But the origins of computers also included ideas and methods that were not about calculation. The idea of a computer program, for instance, originated with punched cards. The first punched cards were introduced to control the fabric patterns produced by textile looms. The first electronic computers, built during World War II, were used to crack secret codes. Computer scientists were very quick to realise that they could use computers for all sorts of manipulations.

The links between computers and language go back to the beginnings of computing. As we saw in the first chapter, computer languages arose from the need to develop complicated procedures that can be repeated at will.

In principle there is no real difference between a computer program that carries out a numerical calculation and one that manipulates text. One involves moving and copying numbers; the other involves moving and copying words. For instance, adding up a shopping list involves taking numbers one at a time from the list and adding them to a running total. Searching for a given term in a list of names involves checking each item in a list for a given value or string pattern. One of the most profound insights of the 'nature as computing' paradigm is that scientific models of nature need not be restricted to mathematical formulae. For example, just as we can use computer languages to write computer programs, so we can use languages to describe processes in nature. To illustrate this, let's look at turtles before we go on to ants.

The turtle as artist

In the late 1970s the Stanford mathematician Seymour Papert devised a system of graphics called *turtle geometry*. The aim was to make it easier for children to learn mathematical ideas by embodying them in concrete form. The basic idea of turtle geometry is to think of a geometric shape as the trail left by an imaginary turtle as it wanders around on a surface. Suppose that we attach a pen to a turtle and place it on a table covered with paper. Then, as the turtle moves around, the pen will leave a mark that shows where the turtle has been. Now, as it moves, the turtle alternately moves forward (or back) some distance and occasionally turns to the right or left. We can at once devise a simple language to describe what happens. To start simply, suppose for a moment that every turn the turtle makes is a right angle. We can use F to denote one step forward, and L and R to denote left and right turns. Then we can describe a trail in terms of the sequence of actions that the turtle makes as it moves, such as FFRFFFLFRFF. Strings of symbols like this form a simple language. Not only can this language describe a path, it also allows us to tell the turtle how to draw a pattern. In the computer programs that implement turtle geometry, the 'turtle' is usually represented by a dot moving around on the computer screen. This dot moves around drawing a pattern in response to any string of actions that we present to it. For example, the string FRFRFRFR tells the turtle to draw a square.

Turtle geometry embodies many mathematical ideas. The idea of an inverse, for instance, is widespread in mathematics: in arithmetic, subtraction is the inverse of addition; in calculus, differentiation is the inverse of integration. Turtle geometry embodies the notion of inverse in a very visible way. For every action that the turtle can take there is an inverse action that returns it to its previous state. The inverse of F (go forward) is B (go back); the inverse of L (turn left) is R (turn right). So the string FB or the string LR each indicates that the turtle immediately undoes the action it takes first. The idea of an inverse extends to any path at all. To obtain the inverse of a given path, we simply reverse the string and replace each term with its inverse. Here is how this procedure works for the example given earlier:

The original path:	FFRFFFLFRFF
Reverse the string:	FFRFLFFFRFF
Replace each term:	BBLBRBBBLBB

So the path FFRFFFLFRFFBBLBRBBBLBB would see the turtle wander a short way and then backtrack to its starting point.

Turtle geometry also embodies many computing ideas. For instance, if we give names to particular paths, we can use those names to denote whole elements within a larger pattern. In drawing (say) the square as described above, we could give the string FR the name 'SIDE'. The whole pattern FRFRFRFR then reduces to '4 SIDE'. However, we can go a lot further than this. If we then give the whole pattern the name 'SQUARE', we can use it as an element in drawing bigger and more complex patterns. Using names for patterns in this way is exactly like giving names to subroutines or functions within a computer program.

Turtle geometry embodies iteration too. For example, we can allow the turtle to make turns at any angle, rather than always 90°. Suppose that we modify the element

94

Figure 5.3 Patterns generated by turtle geometry. The paths shown are described in the text.

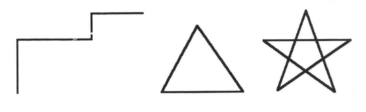

SIDE above to be SIDE(ANGLE), where ANGLE is a number telling the turtle to turn by whatever number of degrees. That is, SIDE(ANGLE) is another name for the string 'F R(ANGLE)', where 'R(ANGLE)' means 'turn right by ANGLE degrees'. If we set ANGLE to be 120 and set the turtle walking, then after drawing three sides its trail starts retracing itself. So the resulting pattern is a triangle. If we repeat this experiment with the value of ANGLE set to 72, then the result is a pentagon. Other angles produce more unusual figures. With ANGLE set at 144, the result is a five-pointed star. With ANGLE set at 95, the result is a rosette pattern.

The point of the above discussion is that turtle geometry is intimately tied to the turtle language. Each expression in the language corresponds to a pattern on the surface. By our introducing more elaborate rules into the language, turtle geometry becomes capable of creating virtually any pattern at all. Let us look at two simple examples. First, we can introduce a growth operator, G, that magically shrinks or enlarges our imaginary turtle. So the command 'SIDE(ANGLE) G(0.9)' would draw a side and then shrink the turtle to 90 per cent of its original size. If we repeat this process, then the result will usually be an inward spiral. If we set the growth factor greater than 1, then the path will spiral outwards. Finally, we can greatly enlarge the ability of the turtle language to describe complex patterns by exploiting the ability to give names to strings. As an example, look at the following set of definitions:

```
SIDE(ANGLE)   =   F R(ANGLE)
CELL          =   6 SIDE(60)
CLUSTER       =   6 (CELL F L(60))
HONEYCOMB     =   CLUSTER SIDE(60)
```

In this set the CELL produces a hexagon, CLUS-TER produces a cluster of hexagons and HONEYCOMB (when repeated) produces a honeycomb pattern. This final example embodies another important idea in computing: modularity. If we were to sit down and try to write out a string to describe the turtle's complete path in drawing the honeycomb pattern, it would be a long and difficult process. However, by breaking it into a hierarchy of modules that produce different levels and elements of the pattern we can approach the problem in a simple and systematic way.

From turtles to foraging ants

How does turtle geometry relate to the behaviour of real animals? Perhaps, instead of turtles, we should look at ants. To move from the simple 'behaviour' embodied in turtle geometry to the behaviour of real animals requires several changes. First we have to consider how real animals move. All of the patterns set out above were deterministic. That is, the path taken was completely predictable from the start. A foraging ant, on the other hand, moves about essentially at random. It does not move according to a fixed pattern. Secondly, and most important of all, an ant interacts with its environment.

Many of its movements are determined by what it encounters as it moves around. If an obstacle blocks its path then it searches for a way around. If danger threatens then it goes into panic mode and tries to escape. If it detects food then it homes in on the scent. Finally, ant behaviour is not solely concerned with movement, but with all manner of activities. Many of the rules that govern an ant's behaviour concern the type of behaviour, rather than the exact details of how it moves.

Another natural phenomenon that is literally computing is the processing of sensory information. Just like a computer, our senses are the input devices of our brain; speech and action provide the outputs. Sensory perception is a matter of selectively throwing away information. Take human vision, for instance. Every second our eyes receive more information than we can possibly process in detail. The miracle of vision is that we are able to distil literally terabytes of sensory data into a manageable volume of information. The process of sight thus involves selective ignoring of most of the details transmitted, so that only vital features remain. What interests us in the world are the shapes and reflectance of surfaces. We're not interested in absolute brightness level or even the actual spectral distribution of light that comes into the eye.

It's a common experience that colours under fluorescent light may look different outside. This is an example of a breakdown of a visual strategy which normally works very well: colour constancy. If the spectral distribution of the light changes (e.g. at sunset, when it shifts to longer wavelengths, or the red end of the spectrum) then the light reflected from a surface changes. But that has nothing to do with the actual surface and it would confuse the recognition process. So we work back from the spectral distribution of light in the early stages of vision to *colour* in a module of the brain referred

97

to as area V4.[9] We've thrown away the specific details of the light's brightness and spectral distribution.

Similarly, snakes, for example, can have trouble seeing freshly killed prey right in front of them and yet are able to detect even the slightest movements in their field of view—their eyes are adapted to select the sorts of information that are vital to their survival. Snakes want to kill live prey rather than eat carrion and can therefore ignore a lot of static pattern information used by other animals. This ensures, of course, that the meat is fresh but in some snakes it serves another function. Snake venom has pre-digestion enzymes that can be usefully circulated in the animal's blood stream before it dies, making the bite from some snakes, such as the African puff adder, particularly nasty even after hospital treatment has avoided death.

How to get up in the morning

When we look at human behaviour, there is good evidence from psychology that humans behave a bit like robots. In particular, we all develop habits. Habits are fixed patterns of action that we follow in a given situation. For instance, one person's habit might be to scratch his or her head while thinking; someone else might bite the end of a pencil while writing. On a higher level, regular activities such as going for a drink after work may be habits too. When we think of habits we tend to think of unusual, and often undesirable, behaviour—such as a child picking its nose. However, habit-formation seems to be linked to important processes in human learning and behaviour. The great French child psychologist Jean Piaget introduced the idea of schemas to help explain mental development in children. Piaget suggested that children learn by developing what he called 'schema', or patterns of recognition, thought and action. A schema is a 'recipe' for dealing with a particular type of situation. For example, if you walk into a room for the first time, you expect to see certain features: walls, ceiling, floor, furniture and so on. The importance of identifying the features that we expect is that once we have identified them we can forget most of the details. In general we need remember only the features that are unique or distinctive. A room with no floor, for instance, would certainly be memorable. As we saw earlier, this selective forgetting assists perception, because we do not need to process so much information.[10]

The above expectations, and the way you react to them, constitute what might be termed an 'entering a room' schema. An important mode of learning is what Piaget called 'assimilation'. We assimilate (i.e. cope with) many new experiences by adapting existing schemas that deal with similar situations. For example, the idea of entering a room generalises to entering all kinds of structures, such as a shed, a tent or a cupboard. Genuinely new experiences (e.g. travelling overseas for the first time) are memorable because we do not have the perceptual framework set up to help us ignore details.

And if you still don't believe that humans are programmed automata, then consider the following question. What happens when you get up and go to work each morning? Most people answer this question with a list something like the following:

> get out of bed
> have a shower
> get dressed
> have breakfast
> pack my briefcase
> leave home
> catch the bus

The details may vary from person to person, and even from day to day. But we can all identify with this sequence of events. Most of us have a routine that we follow on a regular basis. The point is that we are all creatures of habit. The above sequence is essentially a program for getting up and going to work. What is more, each step, each 'routine' in our going-to-work program, has subroutines. Take just one step, say having breakfast. A typical 'subroutine' for having breakfast might be as follows:

get out cereal, milk and spoon
fill bowl with cereal and milk
heat bowl in microwave
sit and eat cereal
clear breakfast table

Each of these steps can again be broken down into a yet finer repertoire of habitual routines that we use. Eating cereal, for instance, might follow a sequence something like this:

dip spoon in bowl
remove spoonful of cereal and milk
put spoon in mouth
chew and swallow cereal
if bowl contains more cereal
then repeat the above steps
or else finish eating

The point of all this is that our behaviour is built up out of schemas piled on schemas. Elementary schemas, such as holding a spoon, are learned in early childhood. In later life we exploit them as givens from which to create more complex behaviour patterns. In the following sections we shall see how new features can emerge out of simple rules of behaviour.

From syntax to synergy

Suppose that you are caught in a blizzard and know that there is a hut at the top of the mountain. How would you find the top? The answer is simple. Just keep heading uphill until you reach the top. This example is a simple case of a 'rule of thumb'. We use such rules constantly in our daily lives. Many well-known sayings are really rules of thumb: 'Never stand under a tree during a thunderstorm'; 'Do unto others as you would have them do unto you'. Likewise, many rules—e.g. 'Keep well away from other cars when driving'—are learned from experience. Rules of thumb are ways of coping with uncertainty. Given any problem to solve, and a complete set of facts, we can usually nut out the best solution. For instance, suppose that you are driving through a strange city; then if you have a street map you can work out the route that will get you to where you want to go. Unfortunately, life is rarely that easy. What do you do if you are trying to find your way through a strange city without a map? Well, of course, you use a simple rule, such as to follow major roads heading in the direction you want to go. The point is that, in the absence of complete facts, we cannot plan a strategy that we can guarantee will 'win' (achieve our desired goal). We may find ourselves caught on a freeway with no exits. We just don't know in advance what will work and what will not. So we adopt a strategy that experience shows will work most of the time.

Next we see that rules of thumb can have profound consequences when applied *en masse.*

101

The bull ant and the bumblebee

The walk home from our university passes through several hectares of bushland. At one point there is a clearing where a road passes through the trees. By the side of this road there is a large ant nest, one of many in the area. The nest is easy to spot: it is marked by a circle of cleared earth about ten metres in diameter. Late in the afternoon the ants are very active. Thousands of them swarm all over the nest. Every shred of litter is at once cleared away by workers. Meanwhile, soldier ants guard two dozen or more entrances. Thousands more ants head off into the bush to forage for food.

The nest gives every sign of being well organised and well designed. If we could look inside, this impression would be even stronger. Eggs and food are stored in separate areas and the nest as a whole is kept clean and tidy.

Ants are not alone in building elaborate homes for themselves. The precision and regularity of a beehive, for example, is even more striking.

How do they do it? That animals with no real intelligence can design and build such well-organised nests has long been a source of wonder.

The answer to this puzzle is simple, but amazing nonetheless. There is no blueprint, no central planning. Ants and bees have no idea what a complete nest should look like. The insects just act according to a very simple repertoire of behaviour. An ant nest or a beehive emerges from myriad simple interactions of individuals with each other and with their surroundings.

To appreciate how an orderly nest can emerge in the absence of an overall design, let's look at how ants sort materials. Suppose that an ant is wandering around the nest and comes across a stray egg. It picks up the egg and carries it away. It has no plan that tells it where to take the egg; it simply carries the egg with it as it wanders

102

Figure 5.4 Formation of a virtual ant colony in the computer program Xantfarm. Instead of grains of sand the ants move screen pixels, creating hills and tunnels.

around the nest. Eventually, in the course of its wandering, it comes across a pile of eggs and drops the egg that it is carrying. Now this pile of eggs may start off simply as a random concentration of two or three eggs. However, as the ants add more and more eggs to the pile, it grows into the nest's central egg repository. The whole process requires only two rules of behaviour to make it work:

- IF you see an egg THEN pick it up
- IF you see a pile of eggs THEN drop the egg you are carrying

In Chapter 2 we saw how foraging and other aspects of ants' social behaviour can be described by equally simple rules.

103

There is an amusing computer program, Xantfarm, that embodies the above ideas.[11] Xantfarm is used as a screen background on Unix workstations. To represent the ground there is a line across the screen, with white sky above and green earth below. Little 'virtual ants' wander back and forth on the ground, as well as up the sides and across the top of the screen. Whenever an ant finds a loose pixel (a virtual grain of sand) it picks it up and deposits it on a pile of sand. Leave the program running for several days and the virtual ants create huge piles of sand with tunnels burrowing into them from the side. In short, they build a virtual ant colony.

Interaction between 'agents' and their environment is an important source of order in many natural systems. Some of the key early work on this topic was done by the team of Paulien Hogeweg and Bruce Hesper at the University of Utrecht. They showed that the structure of many living systems, such as ants or bumblebees, could be explained by the TO DO principle. That is, the insects simply do whatever there is to do at any time and place.

The TO DO principle has many potential applications. Take robotics, for instance. The traditional approach to building robots was to pass a robot's visual input to a large computer, analyse a scene, and then make decisions about action. For years this centralised approach became bogged down. In effect, robots were trying to understand their entire environment before they could act. However, Rodney Brooks, a researcher at MIT, showed that such elaborate interpretation was not necessary.

In his robots, legs and other systems are all independently controlled. Simple rules govern how each leg responds to circumstances at any given time. If a leg is tilted back, for instance, it is programmed to lift and move forward. Feedback systems ensure coordination between the legs. Behaviour of the entire robot emerges out of interactions between all these different systems. No central planning or intelligence is involved.

One result is that robots can now be made smaller and more adaptable than ever before. Some of these robots resemble large insects, not just in their appearance but also in their behaviour. In the future a home may come to resemble an ant colony. Imagine a hundred tiny robots wandering around the house while you are away, cleaning up, killing cockroaches and sorting belongings.

CHAPTER 6

ON THE IMPORTANCE OF BEING WELL CONNECTED

> She has no money, no connections, nothing that can tempt him to—she is lost for ever.
>
> The upstart pretensions of a young woman without family, connections, or fortune. Is this to be endured! But it must not, shall not be.
>
> Jane Austen, *Pride and Prejudice*

In Victorian England, social connections were everything. For any young lady or gentleman in polite society it was essential to be born into the right family, to go to the right school and to join the right club. With good family and social connections came invitations and introductions, good marriages for young ladies and shining careers for young gentlemen. Lacking good connections was a severe handicap when you set out to make your way in the world. Victorian England was not the only society in which social connections were important. They are important in any society. They are still important today.

Just what do we mean by *connections*? Connections are links between people. Besides social links and family links, connections can include just about any way in which people interact or relate to one another. For instance, there are neighbours. There are professional colleagues, clients, customers. There are travellers thrown

106

together by chance on the same train or plane. There are patients in the same hospital ward. Students who study together, people in the same theatre. There are members of clubs, societies, political parties or religions. Many features of society rely on one or another of these connections. In a small town or neighbourhood, for instance, gossip depends on people talking to each other. The spread of disease depends on people coming into close contact with one another. Fads and fashions depend on people influencing each other's likes and dislikes.

THE BALL AND STICK MODEL

At first sight this idea of connections may seem too trivial to be taken seriously. Many scientists have expressed this view. However, the very idea of connections leads to some deep and fundamental insights. It turns out that many ideas that have formerly been studied separately are really consequences of connections. In many respects it is the connections themselves, rather than features peculiar to a specific system, that control the structure or behaviour of a system

Before going any further we need a way to express the idea of connections. One simple approach is the so-called *ball and stick* model. We represent any system as balls (the objects) joined together by sticks (the connections). Now, real people don't look like balls—but remember that this is just a convenience. All we need to capture is the essence of connections. To achieve this it is enough to represent people as balls, and social connections as sticks. In the ball and stick model people and their social behaviour are no different from (say) towns joined by roads or stars tugging each other through gravity. The details of the system do not matter. Only the number and pattern of connections matter.

Mathematicians call the ball and stick model a *graph*. A graph is simply a set of nodes joined by edges. That is, in a graph each ball is called a *node*; each stick is called an *edge*. The ball and stick model is ingrained in our thinking. We draw boxes and arrows on the backs of envelopes to explain many things. Who has never tried to explain, say, the organisation of their company, or their family tree, by drawing names joined by arrows? We also draw electronic circuits and computer flow diagrams as lines joining shapes.

In a very real sense the ball and stick model pervades our thinking. It is even built into our language. Take the sentence:

The dog ate the bone.

If we look at the syntax, we see that this statement takes the form:

subject/verb/object

Both the subject and the object are things (*balls*) and the verb defines a relationship (*stick*) between them. Whether we realise it or not, a lot of science involves identifying ball and stick relationships. For instance, taxonomy is not just about identifying species. It is mostly about identifying how different species (the *balls*) are related (the *sticks*). Many studies are about relationships, such as: *does smoking cause cancer?*

As we have seen, one important insight of the ball and stick model is that many different systems have a deep similarity. The organisation of large corporations has its ball and stick structure in common with the taxonomy of the mammals.

A game of chess

It is beginning to seem as if the ball and stick model applies almost everywhere. And it does. Not only does it refer to how things are put together; it is also relevant to the way they behave, change or grow.

Take a simple example. Suppose that two people are playing chess. We can describe the state of the game at any time by giving the location of each piece on the board. From any given state of play the rules of chess allow for many possible moves. At the very beginning of a game white can choose, for instance, to move any one of the eight pawns or either of the two knights. There are many ways in which white might move a single piece. The king, for instance, might be jumped to the centre of the board. But the rules of chess outlaw this move. Of the 512 ways to move a single piece from the opening state, only 20 are legal.

Now, the important thing to note here is that any move we make connects one state of the board to some other state. Suppose that we could plot all possible states of the chessboard on a map, with each state, each arrangement of the board, represented by a dot. Then we could view an entire game as a path leading from the opening state to one of many possible final states.

When we look up at the sky, or look at a beach, we can at once see how vast are the numbers of stars, or the numbers of grains of sand. Though not immediately obvious, the number of different ways in which a game of chess can proceed is equally vast. To see this let's consider (say) the first ten moves in a game. For the sake of simplicity, suppose that no pieces are taken and that on each move there are 20 possibilities—as on the first move. Then in 10 moves there is a total of 20^{10} or 10^{13} possible sequences of states (i.e. 1 followed by 13 zeroes). However, the paths allowed by the rules of chess are only a small subset of all possible paths within the *state space* of all possible states of the board. If we allow any move of one piece at a time then the number of possible sequences of moves rises to 512^{10} or about 10^{28}. If we place no restrictions at all on changes to the arrangement of pieces on the chessboard, then there are about 4×10^{182} possible sequences of states of pieces on the board. That's 4 followed by 182 zeroes—a truly astronomical number.

What is true of chess is also true of most sorts of changes. For example, at any instant in time an animal moving through a landscape has a location, a speed and a direction. These values define its state at any one time. Its movements define transitions from one state to another.

The ball and stick model also applies to computers. The state of a computer is the net value of all the values in its registers. The program running on the computer decides what new state the current state will link to. So we can describe the behaviour of a computer program in terms of the ball and stick model.

This observation is important because it means that any process that we can model using a computer also has a ball and stick behaviour. It is beginning to look as if we can describe almost anything at all using the ball and stick model. This indeed turns out to be the case.

CASCADES AND CATACLYSMS

We have now seen that the ball and stick model can represent just about any system at all. It applies both to organisation and to behaviour; to forests and to fires; to chains of coral reefs and to nuclear explosions. It is universal.

So what? How does it help us to know that a coral

Figure 6.1 **The connectivity avalanche in a random graph formed by joining balls with sticks.** The diagrams show different stages at which the graph becomes saturated with clumps of increasing size. In each case notice that adding further sticks to join balls within different clumps immediately produces a clump twice the size, so clump sizes increase by doubling: 2, 4, 8 and 16 balls respectively. As the clumps increase in size their number decreases, so it takes only half as long to saturate the graph with larger clumps. This acceleration effect leads to the connectivity avalanche described in the text.

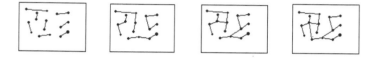

reef is related in an abstract way to a set of balls and sticks? Why should it matter that we can represent the behaviour of a computer in the same way? It is a truism in science that theories of everything explain nothing. On the other hand, it is also true that new ways of looking at things, especially if they relate apparently different things together, often lead to deep insights about nature. And this is the case with the ball and stick model. True, its descriptive power is limited, but at the same time it strips away all the messy details that confuse the issue when we look at different phenomena. It allows us to see the forest for the trees. It shows that beneath all that goes to make a coral reef different from a computer program is a deeper, simpler truth, one that is the same for both systems.

What the ball and stick model achieves is to reveal that what scientists have always studied as separate and

111

unrelated phenomena have a common underlying cause. It allows us to see the real nature of critical processes such as nuclear explosions, epidemics and the onset of chaos. In the early 1960s, two mathematicians—Erdö and Renyi—were studying the properties of large graphs. Now, a *graph* in this case is not a plot of y against x, such as a chart of stockmarket fluctuations. It is simply another name for the ball and stick model. What Erdö and Renyi found was that randomly constructed graphs inevitably contained what they termed a 'unique giant component' when the density of edges (sticks) exceeded a certain level.

To understand what this means, suppose that we have a number of villages in a large forest. At first there are no roads, so transport is difficult. Gradually, however, roads are built between some pairs of villages. As time goes on, some individual villages acquire a second road link. So it becomes possible to walk by road from village A to village B, and then by road again from village B to village C. Eventually, the system of roads becomes so widespread that we can walk by road from one village to any other village at all. The important thing in this example is not that all the villages eventually become linked when there are enough roads. What matters is the way in which the connections spread.

Suppose that for some set of villages A, B, C, . . . we can walk via a series of roads from any one village to any other. That is, there are roads for each pair A-B, B-C, C-D and so on. We can call a set of villages that are joined this way a 'cluster'.

Assuming that roads are built at random, then *at the beginning* each new road will join two villages, neither of which has roads leading elsewhere. So the size of the largest cluster will be just two villages. For a long time no larger cluster will form because the density of roads is just not great enough. Eventually, though, the region will become saturated with roads. At that point it becomes almost inevitable that a new road will link two villages that are already linked by road. So the new cluster is likely to contain four villages. Continuing this process we see that the region begins to fill with clusters of four villages. Moreover, because the new clusters are twice the size, it takes only half as long for the region to become saturated.

At this point a cascade effect occurs. Each time the region becomes saturated with clusters of a certain size, the next stage forms clusters that are twice the size. And because these new clusters are twice the size, it takes only half as long for the region to become saturated again. When still larger clusters start forming, they are again twice the size of the previous ones—four times the size of the earlier ones—and again it takes only half as long to saturate the region. That is, the pace of cluster formation accelerates very quickly. So once larger clusters start forming, they expand rapidly. As the size of the clusters grows, so their number decreases, until one cluster absorbs all the others. This super-cluster appears when one cluster gains a size that includes over half of the villages. From this point onwards, no other cluster can achieve a size equal to this *unique giant component* (UGC), as Erdö and Renyi called it. The process then becomes a matter of absorbing all the remaining clusters into the UGC. The rate of growth of the UGC falls off quickly until it has absorbed all of the villages.

What is true for villages in a forest is also true for any graph. Any time that we randomly add edges (or sticks) to a given set of nodes (or balls), there is a sudden phase change within the graph as a UCG forms. The nodes switch from being mainly disconnected to fully connected. Moreover, the process has some important properties. First of all, the number of edges that we have to add before a UCG forms is predictable. If we

113

have N nodes then we need to insert close to N/2 edges on average before the UCG appears. Secondly, as the number of edges increases, the number of clusters at first increases until saturation point is reached. Then the number declines again as larger clusters absorb smaller ones into a single cluster—the UCG.

Thirdly, the composition of the system becomes inherently unpredictable as the UCG forms. To understand what this means, suppose that we perform an experiment where we fill in a certain number of edges in our graph. Then if we repeat this experiment many, many times we can say that a result is predictable if it comes up most of the time. If the number of edges that we add in the experiment is small, then the state of the graph is highly predictable: the maximum cluster size is nearly always two and most nodes are isolated. Conversely, if we add lots of nodes, then almost all the nodes will belong to the UCG. However, if we add around N/2 edges, the critical number at which the UCG appears, then both the size and the composition of the UCG are extremely variable.

Scientists measure variability using a number called the *standard deviation*. In this case we can use the standard deviation of the size of the largest cluster in the graph as a measure of variability. If we plot this variability against the number of nodes added, we find that its value is nearly zero when the number of edges is either very low or very high. On the other hand, it shoots up dramatically at the phase change where the UCG appears.

Finally, the traversal time for the system is maximum at the phase change. To see what this means, return for a moment to the example of villages in a forest. We can ask what is the longest time it takes to walk from one village to another. Let us refer to this time as the *traversal time*. A simplistic measure of this time is the number of roads we have to follow between villages. At first, when roads are scattered and never link up with others, the traversal time is 1. As the size of the clusters increases the traversal time goes up. When all the villages are finally linked into one giant cluster, the traversal time is large because many detours are necessary. The connections are not as efficient as possible. However, as still more roads are built, shortcuts become possible so the traversal time decreases. Now, taken in isolation, the above findings about connections in graphs are simply a curiosity. But the name *graph* is simply another name for the ball and stick model that we looked at above.

Remember that the ball and stick model is implicit in virtually any complex system. So the phase changes in connectivity that occur in graphs can occur in any complex system. Just what this means depends on what aspect of the real world the ball and stick model represents. To start with, recall the starfish outbreaks we discussed earlier. The ball and stick model is relevant because the balls can represent coral reefs and the sticks can represent the spread of starfish larvae from one reef to another. In this case the reefs are relatively well connected. The series of outbreaks could be interrupted if currents failed to carry the larvae from reef to reef. This might happen if the number of outbreaks were reduced or if the size of an outbreak at some key point was severely dampened. The same sort of property is true of epidemics. The phase changes in graphs reveal the wisdom of isolation—the traditional method of halting epidemics.

CHAPTER 7

PATTERNS IN THE SAND

The Mandelbrot set is one of the most frequently displayed mathematical images. It has appeared in many places, in art, in illustration and in mathematics itself. It is a *fractal*, probably the first to be identified as such to the public at large, and was brought to the fore in Benoit Mandelbrot's famous book, *The Fractal Geometry of Nature*. But the set named after him—the example of a fractal that everybody has heard of—doesn't look like anything from the natural world. Yet Mandelbrot, at IBM, and subsequently with colleagues such as Richard Voss, led the way to our seeing fractals as an excellent representation of many physical and biological patterns; in fact, they are almost the rule rather than the exception. People have been looking at fractals for geological time, so let's look at natural fractals.

UNIVERSALITY OF FRACTALS: EXAMPLES FROM THE REAL WORLD

Bob Mossel, a leading landscape photographer, has for many years been taking photographs of patterns in nature. As a keen amateur pilot, his photographs from the air reveal features and structures not obvious from the ground.

River deltas, sand dunes, salt lakes and many other

Figure 7.1 The Mandelbrot set. The set is actually a map of the border (phase change) between two kinds of behaviour in the formula $X^1 = aX = + b$, for complex numbers. The set contains many regions that resemble the entire set itself.

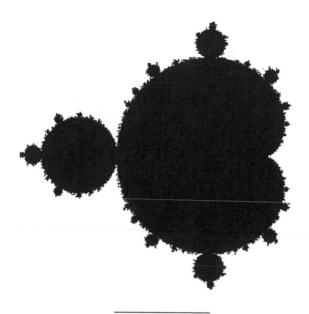

natural patterns are all examples of, on the one hand, self-organisation and, on the other hand, fractals. The process of fractals is endemic, a fundamental part of nature—in plants, sounds, animal markings. In fact they are far more prevalent in nature than the simple recti-linear constructs of man-made objects. Fractals appear everywhere because they are natural growth processes. They are characterised by scaling, endless repetition and growth of simple units, as shown by the example of the Blackwort fern popularised by Michael Barnsley in his book, *Fractals Everywhere*. As well as characterising static

117

patterns, fractals also characterise complex dynamics and the shape of chaotic attractors, which we will discuss further.

THE IDEA OF SELF-SIMILARITY

The basic idea is that of self-similarity. To understand it, let's look at the generation of the branch of a tree. Each branch develops buds which themselves produce new branches—each new branch a tiny replica of the old one. It doesn't have to be exactly the same. This process is captured in the L-systems we saw in Chapter 5. We can add some randomness to get something which looks more like real life, too.

There is one important difference, though, between any fractal model of nature and the actual phenomenon. There is only a finite range of scales for which the approximation is valid in the real world. Real objects don't keep getting smaller and smaller, as does the mathematical entity. Nevertheless, some fractals such as Brownian motion, do scale by factors of over a million. This property of endless repetition down to infinitesimal scales creates the special space-filling properties of fractals. They are almost like solids, but not quite—more like very fine sponges—giving rise to one of their special, well-known properties, that of fractal dimension. These are tricky mathematical concepts. Intuitively, there are many analogies in the natural world. Foam is a good one: it can occupy a huge volume but will condense down to just a tiny amount of liquid. Talcum powder is another: a fine dust which will cover a huge surface, and itself having a huge surface area.

Our childhood introduction to dimension is a very simple one. There are three spatial dimensions, with points, lines and solids having one, two and three dimensions. Later we might start to think of time (a

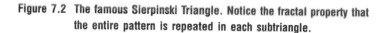

Figure 7.2 The famous Sierpinski Triangle. Notice the fractal property that the entire pattern is repeated in each subtriangle.

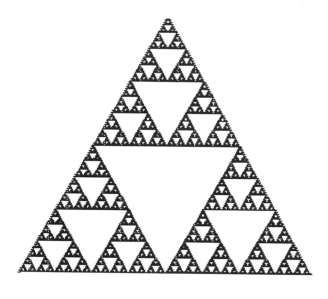

closely related quantity) as a further dimension. Fractals take us into the new territory of *non-integer dimensionality*.

The oft-quoted example is the length of a coastline. Suppose we were to start in Ullapool at the far north of Scotland and travel by small boat down the west coast of Britain to Penzance in Cornwall. We would, all being well, make it down there in a reasonable time; and if we had travelled at constant speed we could work out the length of the coastline by dividing the time we took by our speed. The coastline of Britain, however, is pretty ragged, particularly the west coast of Scotland. If we were to go into and around every loch, hugging the coastline very tightly, it would take us a lot, lot longer.

119

But our boat will be simply too big to follow the line of every tiny inlet. If we want to do that, we'll need a smaller boat. It should be easy to see, though, that as our boat gets smaller and smaller and the coast gets more and more jagged, we are never actually going to get to Penzance. So even though we have been travelling along a line, we're really attempting to cover an infinite distance.

Over the last hundred years, mathematicians have had a lot of fun with infinity. Not all infinities are the same size, according to their subtle arguments. The paradoxical features of fractals, infinite but not quite filling space, are examples of this hierarchy of the infinite.

Dynamic spatial patterns exhibit self-similarity too. Waves on the sea are such an example. Tour operators love to advertise small boats sailing in pictures of an island paradise. What could be more relaxing than lying out on the trapeze of a catamaran lazily sauntering around in the sunshine? The truth is often not so simple. If it's windy enough to lie out on the trapeze, there are also likely to be waves. Going upwind (at a slight angle, since you can't sail directly into the wind) causes the most trouble. As the waves hit, your small boat gets thrown around and you need to keep your wits about you to avoid falling off. Now, if all the waves are the same size a gentle rhythm is still possible, but they are not, and every so often a large wave will come along which will knock you off the boat. Waves come in different sizes. But although there seems to be a self-similar property here, mathematical models of wave systems are not necessarily fractal: thus some caution is necessary in assuming that every system with a range of scales is actually fractal.

Surfers also talk of big and small waves and trying to guess when the next massive wave will hit. Imagine now that a storm comes through and your island paradise becomes somewhat less idyllic. The wind churns up the sea and, particularly around the land, the waves end up coming from all directions. Now you are buffeted this way and that, just like a dust particle in the atmosphere being jolted this way and that by air molecules. Such a simple system hides a very interesting phenomenon, Brownian motion. The trouble with having a name like Brown is that, if you become really famous, there are other famous Browns to contend with. But as a scientist, you couldn't be much prouder than to be associated with the phenomenon of Brownian motion.

In 1827 Robert Brown observed this jitterbug behaviour in pollen particles in the liquid on a microscope slide. It arises through the random blows of air or liquid molecules exchanging energy with the particle so that it itself has the same average energy as the gas molecules. But the jittering looks the same at different spatial and temporal scales. Brownian motion is a standard part of statistical mechanics in most physics courses from school onward. But it has recently become of great interest because of this very property of self-similarity across different scales: Brownian motion is a natural fractal. If we plot positions of a Brownian motion particle as a function of time and space, the traces all look basically the same. But, unlike the pure fractals of the Mandelbrot set, there are wide variations and the traces are not exactly self-similar. The similarity is more *stochastic*: the traces look the same on average, this property being referred to as self-affinity.

BIG WAVES AND LITTLE WAVES

To go a little further into characterising Brownian motion, we need to understand the notion of the power spectrum of a random process. First we need the idea

Figure 7.3 Fourier analysis of square waves

---

of *Fourier analysis*, a numerical procedure for the breaking up of a signal into components.

The idea of waves and frequency is commonplace. We're familiar with waves on the sea, sound waves or even wave-like patterns in the sand. Waves are periodic phenomena, in which the up and down motion, or amplitude, varies in time or space. Just as a big sea will have a very deep swell or very large waves, the energy of a wave depends on the amplitude—the bigger it is the larger the energy. In fact the energy varies as the square of the amplitude, making large waves proportionately more energetic than smaller ones.

The brilliant insight of the French mathematician, Fourier, was that the mathematics of waves has much wider applicability than those systems that have an obvious periodic character. Leaving aside a few mathematical niceties, it is possible to build up almost any shape out of a series of simple waves, even if the shape has sharp edges. This seems counter-intuitive, but examination of Figure 7.3 shows how this happens for the example of a square wave. To get the sharp edges, though, we do need a rather large (in fact infinite) number of components. The simple waves we use here have a particular mathematical form: as shown in Figure 7.3, they are trigonometric functions and are often

122

referred to as sinusoids. A wave retains its shape as we change the frequency, scaling in just the way that a fractal scales. So it isn't surprising that fractals have a simple form in the Fourier domain.

With random or quasi-random processes, it is convenient to deal not with the Fourier analysis but with a closely related concept, the power spectrum. This tells us how much energy there is in each frequency component and is closely related to both Fourier analysis and the *correlation* in the process itself.

Linear systems, which we come to shortly, have special properties which make the wave analysis extremely useful. For example, the blurring of an image is simply a reduction in amplitude of the sine waves of very short period (high frequency) that contribute the fine detail to the image.

Fourier analysis gives us a neat recipe for making natural-looking fractals: the wave amplitudes all seem to decrease as the wave frequency gets greater in a similar way. All we need is that the Fourier spectrum fit within a particular range of shapes. Pure Brownian motion has a frequency spectrum that falls off as the square of the frequency, $1/f^2$, but models of Brownian motion often extend the frequency exponent to an *envelope* of values on either side.

Many natural phenomena have a physical structure in which the amplitude of the pattern falls off as frequency increases. That is, that size is related to $1/f$, the inverse of the frequency f. If we increase the exponent—that is, $1/f^3$, $1/f^4$ and so on—then the amplitude falls off even faster as the frequency increases. Thinking back to the building up of a square wave, we can see that the high frequencies, which change very rapidly, are what produce the sharp edges. So it is with different fractals. If the exponent is too high, then the results look too smooth to be natural. At the other extreme, they look too sharp and spiky.

We shall see later that there are models that underlie fractal patterns which helps to explain why they are so ubiquitous. But there is one very prominent fractal character that has proved much more difficult to explain. These are fractals whose power spectrums fall off as $1/f$. This is at the low frequency end of the range we saw for fractal Brownian motion, meaning that the patterns have more high frequency energy and hence seem sharper.

Examples abound. In physical processes they range from noise in semi-conductors to water levels of the Nile, from traffic on freeways to music. Across cultures and centuries, musical melody has this spectrum. The latest in musak is a fractal, synthesised on the fly, while fractal-derived music also forms part of serious artistic endeavours.

Common mechanisms behind natural fractals

Observing fractals in many natural systems leads one to suspect that similar underlying processes cause them. Remember that a fractal pattern is a global phenomenon. The aerial image of a particular river delta, for example, with its fractal branching pattern, is not the result of an observer tailoring this image. At the local level, as the river carves out its path, there is no knowledge of the big picture. Certain sorts of growth process simply generate interesting global patterns. One such process is *diffusion-limited aggregation*, or DLA, a ubiquitous and much studied process.

To understand DLA, we start with a growth centre in an environment or medium. Particles diffusing past are captured if they come within some given distance. They enter one by one at the outer circle, some distance

Figure 7.4 Examples of diffusion limited aggregation (DLA). On the left is an example of dendritic growth. On the right are structures formed by aggregation of falling particles subject to different levels of Brownian motion and different degrees of 'stickiness'.

from the growth centre. They then move at random along the grid, going in any direction. Sometimes they go back outside the circle and are carried away. Others hit the growth centre, now a growing aggregation, and stick. The resulting dendritic growth has a fractal dimension of around 1.7, similar to phenomena in physics, biology and even the development of cities.

To explain the precise structure of the growth region in qualitative terms is difficult, but we can get some insight as to why a somewhat irregular mass develops. Imagine that we shoot particles towards the growth region from any direction, but in a straight line. If their starting positions are random, we will just get a disc growing out with rough edges. But now, as a particle staggers from side to side, instead of moving in a straight line, it sweeps out a cone and is likely to hit that part of the growing cluster sticking out furthest that lies within this cone. Hence the bit that is furthest out will grow faster

125

than the remainder. But because of the inherent randomness of the particle's 'walk', growth occurs elsewhere too. This intuitive approach shows us that the randomness in the walk (size of the cone) will affect the compactness of the structure, and that if the particles do not stick but have only some probability of sticking, different forms will occur. Thus within this simple framework a large range of patterns can arise.

Although we think of man-made phenomena as being composed of straight lines and regular shapes, seemingly random interactions between people produce other patterns. Studies of cities have shown them to have some fractal shape or morphology. It's easy to see this as DLA: the city grows as people drift in and settle. They are most likely to settle where there is already some habitation. This of course ignores the expansion of the city within any specific controls exercised over development. Nevertheless, the results fit many older cities, as described by researchers Michael Batty and Paul Longley in their book *Fractal Cities*.

Similar mathematics describes the very opposite process to DLA. It is usually referred to as *dielectric breakdown*. The branching structures we have noted might remind you of lightning. Rightly so, since the forked structure of lightning strikes *is* fractal in character. Great charges of static electricity build up on clouds—to the point where the resistance of the air breaks down, a huge discharge occurs and the charge flows to ground. But once the breakdown has started, many new breakdown points arise along the way, giving rise to the forks. The material (air) and the nature of the field (clouds etc.) govern just how the discharge takes place, and again we can get a wide range of patterns from this simple process. So here we have the opposite model for the development of a city: expansion pressure from a single point. Still a fractal phenomenon, but not a cause of DLA.

Thus the goals of complexity research are achieved: models of physical processes, the underlying mathematics, extending to patterns of social development!

A new tool for engineers and scientists

Dot puzzles are very popular with children—join the dots to make a well-known shape—but deciding which dot to join to which involves powerful pattern recognition mechanisms. If the dots are not evenly spaced or are crowded together, the interpolation of the contour in the gaps may be quite difficult. Our sensory systems are always looking for patterns—the pictures in the fire and ink blots mentioned above, for example. Our mistaken perception in visual illusions of all kinds arises because we try to make sense of image data, even when it's somewhat unusual. With fractals so pervasive in the natural world, one might expect that we make use of them for pattern recognition. Although it seems that we are fairly good at recognising spatial transformations of objects, we are rather limited to common movements of solid objects in 3D space. When we classify a pattern we note the elements, but how precise is our recognition of the interrelationships? Would we notice deviations from self-similarity in a fractal structure? At present it seems that we would not, which is quite surprising.

Interpolation is the filling in of the gaps between data or sample points. Traditionally, the methods interpolation have emphasised smoothness, making sure that nothing too wild happens, an essentially conservative approach—and that is what we would normally do in our child's puzzle. But fractals are far from smooth. So if we know or have strong reason to believe that a set of sample points is drawn from a fractal, then the interpolation

127

function should be quite different and should not be smooth. This is a novel area, with much work to be done in it.

Another way to think of this is in terms of the shortest descriptive computer program needed, along the lines of Kolmogorov complexity discussed in Chapter 2. If we have a ragged set of data points, defining a contour through them will be very complex with many line segments to list. But if we know that the raggedness is part of an underlying fractal process, we can get by with a much simpler description. This same approach leads us to think of new and much more powerful ways of compressing data. If we have objects of low Kolmogorov complexity, it makes more sense to recreate them from scratch rather than to send copies of the images.

A fractal is a complex structure, but it is often made from very simple generating rules. Thus instead of sending somebody a copy of a fractal image it takes much less bandwidth or storage to send the generating program. On the other hand, we do not get something for nothing. We have to put a considerable amount of work into designing the program in the first place. Michael Barnsley specialises in the fitting of fractals to images. His techniques, described in his excellent book *Fractals Everywhere*, apply even to pictures one would not at first think were fractals. The method rests on the Collage theorem, which captures neatly the idea of self-similarity. We start with a fractal object: Barnsley's well-known example is a fern. We now look for a geometrical transformation that will shrink and maybe distort the original image to produce a tile. We now try to make a set of such tiles that cover the original picture: the more exact the coverage, the more exact the approximation. Now the set of transformations form an iterated function system, which we can use to regenerate the original image.

Back to the travelling salesman

In Chapter 2 we discussed the Travelling Salesman Problem, highlighting the difficulty of solving the problem for large numbers of cities. But much effort has gone into finding simple approximate methods. One interesting example is the use of a fractal, the Peano curve. The fractal dimension of this curve is 2; that is, it fills the entire circumscribed area. The approximation is as follows. Each city will lie on a particular point on the curve. Now sort the cities according to where they are and that will give an approximation to the shortest route. To see how this works intuitively without a mathematical proof is not easy. At any point the Peano curve grows outwards like a locust swarm devouring every point in its path. Points close together on the Peano curve are always close together in normal coordinate space. So as it sweeps outwards, it collects the cities more or less in increasing distance from the start. But if we were to order the cities according to concentric circles, we would end up with errors where two points close together along a radial line were visited in different sweeps: the wriggling path of the fractal makes sure they all get carried along at the right time. Does this theoretical method have any relevance to the practical world? There is no obvious way in which cities adapt, so it is hard to see any sort of hidden evolution of cities: we have only the specific planning experience of city architects. But it remains the case that a road or transportation network similar to the Peano curve will facilitate delivery systems by providing natural optimisation of travel times.

Since Benoit Mandelbrot wrote his famous book on fractals, we have seen their appearance throughout nature. From complicated shapes and patterns we now turn to complicated *dynamical* processes, but we shall see fractals turn up again there too.

129

Figure 7.5 Formation of Peano's space-filling curve. The first diagram shows the basic pattern. Subsequent iterations repeat the basic pattern at ever finer scales.

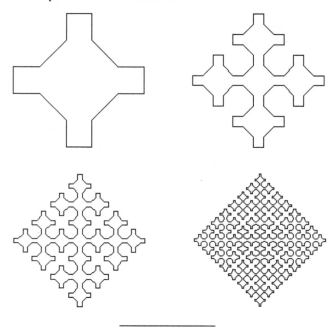

THE CHAOTIC ROLLER COASTER

In the 1980s, *chaos* became a fashion, a household world. Not the chaos of everyday life, though, but a property of dynamical systems. The strange patterns of motion, and the notion that systems could be unpredictable even if there were no noise, captured the public imagination. Chaos became a topic for conversation among sociologists just as much as among ecologists.

Before we can appreciate the nature of chaos, it helps to understand a little of how simple, linear systems behave. Two central ideas in linear dynamics are stability and equilibrium. Some simple experiments may help the reader to understand what these terms mean. First, have you ever tried to balance a pen on its end? The pen needs to have a flat end, otherwise the task is impossible. Anyway, once you have succeeded in balancing the pen on its end you have achieved an equilibrium. Once it is in position the pen will quite happily stay where it is. Now try nudging the pen near its tip, ever so slightly. The pen immediately falls over. The balancing pen is an example of an *unstable* equilibrium. Any shift away from the equilibrium position *increases*, rather than decreases.

Next, instead of balancing the pen on its end, try dangling it on the end of a piece of string. If you suspend it vertically then the pen stays quite happily where it is. Again, it is in equilibrium. Now try pulling the pen slightly to one side and let it go. Immediately it starts swinging back and forth. However, it eventually comes back to rest in the vertical position. This vertical, resting position is a *stable* equilibrium. Any disturbance is soon erased by the system.

Non-linearity often arises through interactions. For instance, a simple pendulum behaves in linear fashion and its motion is predictable. But a certain popular toy, in which a pair of pendulums are attached to opposite ends of a rotating arm, behaves in a highly irregular and unpredictable fashion. In short, it is chaotic.

Non-linear systems differ from linear systems in several important ways. One difference is that linear systems tend to obliterate minor variations whereas non-linear systems magnify them. Consider the following examples. You are taking a bath and splash the water. This action makes a lot of movement on the water surface and makes the water depth very uneven. But after just a minute or so the water surface settles down and becomes flat and even again. Suppose that the water becomes cold and you add hot water from the tap. Then, as you add the hot water, the water temperature becomes very uneven. However, even if you do not stir the water it will soon reach an even temperature everywhere.

Both of these examples are cases where physical processes tend to obliterate local differences. They drive the level and temperature of the bath water towards an equilibrium. When the bath is in equilibrium it stays there unless you disturb it, such as by splashing, or adding hot water. In contrast, imagine a huge gas cloud out in space. If the gas is dense enough then the cloud will not dissipate out into space. Unlike the bath water it does not settle into an equilibrium. Instead it shrinks, because gravity draws the molecules together. As it coalesces there are minor variations in density. Instead of becoming smoothed out these irregularities contain centres of attraction that draw the gas inwards. Eventually these centres become stars and planets, and give rise to all the variety of objects that we see around us.

Examples of processes that magnify minor differences abound in biology. In the eye, for instance, nerve cells not only fire messages when stimulated by light but sometimes inhibit the response of neighbouring cells. The result is the tuning of cells, say to edges in an image. During early foetal development and growth, neighbouring cells in the embryo can give rise to completely different organs.

In all of the examples the enhancement of variations emerges out of countless interactions between simple objects. These objects are gas molecules and dust particles in the first example, and cells in the second. Interactions give rise to non-linearity. In general, it is non-linear processes that tend to enhance irregularities. This point has been emphasised by the Nobel Prize winner Ilya Prigogine. He examined the properties of 'dissipative systems' that are far from equilibrium and stressed that the potential for minor irregularities to grow into large-scale patterns is an important general property of these systems.

Irregularities grow into large-scale patterns because non-linear systems are sensitive to initial conditions. This effect was discovered by Edward Lorenz, when he tried to simulate weather processes on an early computer which we describe in more detail in Chapter 7.

It is this sensitivity to initial conditions that makes chaotic systems inherently unpredictable. This is obvious when you think about it. The essential problem is that we can never measure any variable with complete accuracy. Nor can we calculate the value of any parameter with absolute precision. These small errors are enough to ensure that the true behaviour of a system will soon deviate from any model that we use to try to predict its behaviour. So long-term weather forecasting is just not possible. Nor is it possible to make accurate long-term forecasts of (say) stockmarket prices, or about any other chaotic system. Several biologists have speculated that chaos is an important source of novelty in living systems. The point is that minor, local differences can lead to unpredictable, large-scale changes from the status quo.

PHASE SPACE

To understand the interesting patterns produced by chaotic systems, we have to take a special perspective, called the phase-space portrait. In our sketches of Brownian motion, we plotted the position of the particle against time. We could, however, determine its position and momentum at every moment and plot these against one another, with each point on the trajectory representing a particular time. This combination of velocities and positions at any point in time completely describes a dynamic system. It is this phase-space portrait where we can see visually the interesting things that happen in chaotic systems.

Just as we see the pattern created by a spider's web by looking at it across space as a temporal history, so it can be useful to view many complex systems as a pattern—but not just a pattern in real space. In an ideal gas we have many molecules travelling at different speeds. To characterise each molecule at any given time we need to know where it is and where and how fast it is going. It is convenient to build a representation using both these quantities—the position and velocity vectors. If we consider a rarefied gas with just one molecule then we can plot a trajectory in *phase space* as a function of time. This path will be a series of straight lines parallel to the position axes (because the velocity is not changing), with sharp turns as the velocity changes at a collision with a wall.

We can now add more and more molecules, producing a cloud which is a sort of hyper-cylinder along the position axes but with velocity concentrated in a particular band with fuzzy edges. But for many molecules it is useful to think of the phase space as composed of separate axes for each and every molecule. Thus for N molecules we have $6N$ axes. For a gram of hydrogen, N is of the order of 1023, making this rather a large graph. Each state of the gas, every set of positions and velocities, has its own point on the graph, so we can again plot a trajectory of the whole gas as time goes on. This is a very complicated path. It's a thick knot and if we wait long enough it will fill a huge volume of phase space, although it will take rather a long time to get out to all the velocity extremes.

This random trajectory is actually not that interesting. But in some dynamical systems this trajectory develops some special structure, enabling us to describe it in a simpler fashion. It is these systems that are of interest to us in this book. In some cases the simplification is dramatic, resulting in a military-like ordering of all the elements of the system. In others, the trajectory still looks random, but has some special mathematical properties.

This phase space view is valuable for understanding chaos and many aspects of complexity. But for some of the biological systems we will consider it's somewhat limiting because the number of agents (particles in the gas, for example) is continually changing. This is the frontier: observable, but so far intractable.

Attractors and patterns

Dynamic systems have three types of attractor:

1 *A fixed point* The system reaches a point in phase space and does not progress any further. It remains at the same point in phase space until the end of time.

2 *A periodic orbit* The system enters an orbit, visiting the same set of points in phase space over and over again.

3 *A strange or chaotic attractor* Here the system, like the Flying Dutchman, is compelled to wander forever in phase space, never returning to the same point. The patterns traced out in phase space are visually intriguing and have stimulated artists as much as scientists. However, the attractor is very definitely different from a completely random system, which would visit every point in phase space. It is restricted to a limited volume. Furthermore, this volume has limited dimensionality. The attractor is a fractal.

Figure 7.6 Attractors for logistic growth

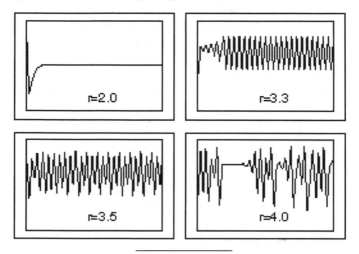

Free will returns!

Why is chaos so interesting?

Everybody, especially the English, know that weather is chaotic. What, then, was so special about Lorenz's discovery of chaos in weather systems? Like many discoveries, it lay fallow for some years, but chaos was to sweep way outside obscure academic journals to become the source of many popular books. Not only was chaos totally different, not only did it generate new and interesting graphics, but it also struck at a much deeper level. Chaos attacked the very heart of determinism and opened up new opportunities for free will!

Before the chaos boom we imagined that, if a system was free of random elements, then, if we knew the starting conditions, we could predict its evolution for ever more. That is still essentially true, but we now know that the specification of the initial conditions has to be impossibly accurate in the case of chaotic systems.

Figure 7.7 The Lorenz attractor. The lines show changes in a model weather system through time. This strange attractor is actually a three-dimensional pattern. Notice that the lines never duplicate the same path twice.

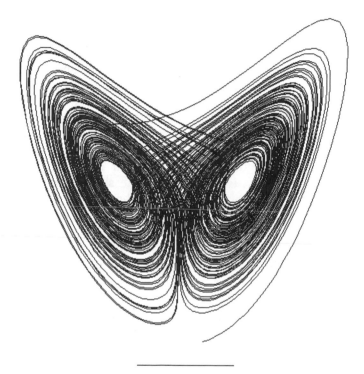

The weather

Although the famous nineteenth-century mathematician, Poincaré, had a deep understanding of many of the non-linear phenomena that make up chaos theory, the start of the initial wave of interest is usually considered to be the meteorologist, Edward Lorenz, at MIT. It isn't really surprising that the weather should be hard to predict: land surfaces are irregular in shape, mountain ranges are far from easy geometric shapes, and water surfaces vary in size, shape and position without obvious pattern. But it isn't this cluttered nature of the world which interested Lorenz. Weather models in today's meteorological offices have many variables, easily as many as a million: Lorenz's model had just three.

The weather patterns on the earth have two main driving influences. Air is heated at the equator and moves out to the poles, from where cold air returns, and the rotation of the earth creates prevailing westerly winds. The earth's rotation has another effect, through the so-called Coriolis force,[12] creating eddies in the atmosphere as the hot air flows into the higher latitudes. It is this large-scale movement of air around the planet that Lorenz, like all good physicists, tried to capture in the simplest possible model. His phase space had just three variables and contained chaotic attractors: orbits in phase space which would travel on and on without rest. Two attractors that started close together would diverge rapidly.

The curious shape of the Lorenz attractor is reminiscent of a butterfly and perhaps for this reason the sensitivity to initial conditions came to be known as the 'butterfly effect'. But there was also another idea lying underneath the surface. The extreme sensitivity to initial conditions of chaotic systems means that the very slightest change in those conditions can produce radically different results. Such a tiny change might be evoked by a butterfly landing—an infinitesimal effect.

THE SCIENTIFIC BLINDSPOT

Although science proceeds by rigorous testing of theory, sometimes its progress slows down for one reason or another. Stephen Jay Gould, in numerous highly readable books, documents the erratic course of science towards better paradigms. Sometimes the blockages are social (such as political pressure which propped up Lamarckian evolutionary theory in Russia), sometimes technical.

In this chapter we've seen how technical advances in computers have brought about breakthroughs in how we look at dynamics and how we describe the objects of the real world. By not trying to make things linear, so that one can do the calculations, nor trying to find simple smooth functions to describe shapes, we have found untold richness of dynamics and form.

In the nineteenth century, pure mathematician Georg Cantor became interested in sets that are like fine dusts, the predecessor to fractal theory. Early in the twentieth century, mathematicians studied sets from mathematical curiosity that we now recognise as fractals. These pure abstract structures underlie patterns and dynamics of the natural world.

Science tends to be a little bit like the drunk who is looking for his wallet under a street lamp at night. When a stranger comes up to help and asks where he lost the wallet the drunk explains that he lost it further down the street. He's searching under the street light because it's the only place where he can see.

The point of this story here is that a lot of scientific research is driven, or limited, by the tools available. From Galileo turning a telescope on the heavens, to DNA recombination, new analytic methods in science have always led to new discoveries. One reason why science did not deal with complexity long ago was that an essential research tool—the electronic computer—was not available. Even today there are some problems that the fastest supercomputers cannot tackle adequately.

CHAPTER 8

IMPLICATIONS FOR COMPUTERS

In the early 1990s, in the border zone of Italy and Austria, a corpse was discovered, not a climber of our time but a fully clothed traveller from the Iron Age, who died thousands of years ago. Imagine the excitement of anthropologists at such a discovery—tangible evidence of what life was like at that time. What would be the equivalent momentous discovery in computing? The idea of machines that think is now somewhat old: machines do think in limited domains; they play chess better than most of us; they can find correlations in huge databases beyond the grasp of any human mind. But could they ever be conscious? Will we ever see HAL, the computer of the movie *2001*?

Despite the advances in our understanding of neural networks, many people still believe that consciousness will never be a property of non-biological systems. Philosophical argument continues to rage, but as technology improves we can ask more and more precise questions. So let's take a look now at biological computation, or biocomputing.

THE HOLISTIC APPROACH TO THE BRAIN

Animal brains don't look a lot like computers. But are the underlying computational principles the same? This

profound question has occupied neuroscientists and philosophers for decades. At the close of the millennium,
controversy still rages on whether we should look at the
brain nerve cell by nerve cell or as a single large adaptive
system. Somewhat less influential at present are those,
like Roger Penrose whom we've already mentioned, who
believe that classical mechanics is not sufficient to model
an animal brain.

The human brain contains approximately 10^{10} nerve
cells, or neurons, each with up to 10^5 connections to
other neurons. It's a giant parallel computer. For something like 30 years, the model of neural computation was
a simple one: the signals coming into a neuron from
other neurons were added together; if they exceeded
some threshold, the neuron would fire a voltage spike to
some other set of neurons (often including some of the
input neurons). So the number of spikes fired per second
would depend on the average number of spikes coming
in. Huge networks take in data from the senses and
ultimately fire spikes to muscles controlling movement of
the body. In the human brain, of course, a lot of activity
can circulate within before any muscular output appears.

Emergent properties of neural networks have come
to the fore in the last few years but as an idea it's quite
old. In 1977 Walter Freeman published a book entitled
Mass Action in the Nervous System in which he argued
that important properties of the brain are captured in
the oscillatory behaviour of populations of cells. In
particular, sets of connected oscillators will phase-lock
under the right conditions, as we saw in Chapter 3.

This view of the operation of a brain was and still is
somewhat radical. The conventional wisdom held that the
activity of a nerve cell increased as its input increased. The
activity is measured in terms of the firing rate of the cell,
the number of voltage spikes fired per second. A vast
amount of neurophysiology and theoretical modelling has

been devoted to understanding single cell activity. We now know in great detail the properties of cells that travel from the optic nerve to the brain; we know there are groups of cells that respond to oriented lines, colour, particular directions of movements, and even faces. It took rather longer for the scientific community to accept that these cells deep inside the part of the brain that handles vision were indeed sensitive to *faces* and not just oval shapes or other primitive figures. But a decade of experimental work has gradually turned the critics around.

The activity of a single cell is a discrete, stochastic signal. But, looking at a population of cells, the average will be much smoother, and with appropriate connections between populations the average activity will oscillate. Collective activity appears as phase-locking of these oscillations. Freeman saw phase-locking as a *major* element in the activity of a brain. His theoretical ideas were derived from experimental work on the olfactory bulb of the rabbit.

A more recent discovery, made by a former student of Freeman's, Charlie Gray, attracted a lot of interest in this rather quiet area. Gray and others looked at the firing rates of cells in the areas of the brain where visual information is first processed, the striate cortex. There are numerous different types of cell in this area. The kinds Gray investigated each look out into their own particular region of space and respond when a line at a particular orientation appears in that region. Looking at the behaviour of just a single cell shows a noisy signal varying roughly linearly with the contrast of a line. But oscillations in the firing rate also occurred when long lines were presented to the visual field, exciting numerous such cells. Even more interesting, these oscillations in the firing rate were found to be phase-locked. This phase-locking could even extend to cells in either hemisphere of the brain. This discovery immediately suggests that the phase-locking, as a collective property, is the way in which the brain encodes an extended line or border.

This result was potentially big news: it is easy enough to record from a cell and deduce that it signals some simple pattern like an edge or a patch of red. But how are more complex signals represented? It's easy to see that the number of ways we can combine primitive tokens together must get far bigger than even the vast number of cells of a biological brain. The information we store is far bigger than the number of cells, just as the DNA of an animal does not in any sense describe the grown animal. So here was an exciting alternative: these complex structures are not represented by single cells, say a long twisty line cell, but by a group of cells along the twisty line which are all operating in synchrony with one another—jumping up and down together in a long twisty line. At the time of writing these issues are still hotly debated.

The coupling of the brain across the two hemispheres is interesting in another way. The two halves of the brain are just the top of a complex hierarchy of modular structures. Throughout the book we have seen ways in which complexity is brought under control by building hierarchies, in computer programs and in life. The brain is far from a huge homogeneous network.

If we look at the human cortex we find it is made up of perhaps a hundred such modules or cortical areas. At this modular division computer science meets evolution. In software engineering and in the development of computer systems generally, breaking up tasks into separate independent modules with defined interfaces between them is the key to managing complexity. Yet the brain clearly derived its modular structure during evolution: new modules developed as brains became more complex. Thus if the brain just evolved that way, it would be hard to say that it has a modular structure for reasons of efficiency. The real situation is undoubtedly a complex mixture of the two.

There is abundant evidence that the brain's modular structure is genetically laid down. As the infant's brain develops there are critical periods in which some structures will grow or not grow depending on the environment. There is simply not enough DNA to 'order' the complete layout of a brain, but there is some complex mixture of direction of growth and experience laid down.

A common everyday experience of this is the learning of a second language. Beyond a certain age learning a correct foreign accent just seems to be impossible. In his highly informative book, *The Language Instinct,* Stephen Pinker points out that Henry Kissinger, former US Secretary of State and a man of undoubted intelligence, retained a strong accent, whereas his slightly younger brother speaks American English flawlessly. Joseph Conrad, a major novelist writing in English, learned the language at the age of twenty-five and, despite his literary genius, retained a heavy accent throughout life.

Like many profound ideas, the idea of innate knowledge is not new. Noam Chomsky caused a furore in the 1960s by propounding just such a suggestion:

> If the conclusions of this research are anywhere near correct, then humans must be endowed with a very rich and explicit set of mental attributes that determine a specific form of language on the basis of very slight and rather degenerate data.

In fact Stephen Pinker describes striking empirical evidence some three decades later, with the discovery of a family with a defective language gene. Its carriers are unable to develop some simple grammatical constructs.

144

Exciting though this is, the numbers of individuals encountered is small, suggesting that we should be cautious in relying too heavily on these data.

Thirty years later we know a great deal more about neural systems and learning theory, but natural language processing is still some way in the future. These questions of innate knowledge and consciousness are the very frontier of neuroscience.

We discussed modularity and object-oriented computation in Chapter 2: in a brain we see the ultimate hierarchy of computational modules. So, having looked briefly at the relatively recent holistic discoveries, let's now dig a little deeper into the nature and action of neurons themselves.

BIOCOMPUTING

In the 1960s artificial intelligence was strongly oriented to rule systems and inference methods. There were some notable successes in (say) expert systems, but in other areas such as pattern recognition and robotics progress was much slower. It always seemed unsatisfactory that things a tiny insect could do should defeat the top supercomputers. It was frustrating to engineers to find that sharks have evolved a *rough* skin surface to *reduce* drag in the water. Thus it was not surprising that eventually people should turn to studying how animals and ecosystems 'compute'. In the late 1980s the trickle of engineers and scientists moving away from rule-based systems, so-called strong AI (artificial intelligence), and into biocomputing became a torrent.

Neural computation

A late twentieth-century computer chip will have something of the order of ten million transistors. Each

145

transistor has just a few input connections. A human brain has 100 to 1000 times more components, each with 10 000 connections, and with much more sophisticated processing of all the inputs. Yet universal computation as defined by Church and by Turing is possible with just simple digital circuits. There are still plenty of people who believe that digital computation cannot emulate the computation of a human brain. This may be true: but we still do not know anywhere near enough about the computation of multi-agent systems to be sure that mental phenomena do not fit into this classical picture. But despite the immense sophistication of today's computers, we have a simple model, the Turing Machine, of computation. We do not as yet have such a definitive model for neural networks, although there are many suggestions.

Hodgkin and Huxley received the Nobel Prize for elucidating the basic principles of a nerve cell, or neuron. It is a very sophisticated system involving several different ions (principally sodium, potassium and calcium) travelling through gates in cell membranes. The gates can be opened and closed while ion pumps in the cell membranes will move ions across, against a voltage gradient. More recently, Penrose has brought increased attention to bear on the detailed, so-called microtubule structure of a membrane itself and the computation (possibly very significant) going on there.

Modelling nerve cells at such a level of detail does not allow one to study large networks, not with present computing resources. So this lowest level of computation is approximated in many biological studies and almost completely ignored in artificial neural networks.

It is difficult to know where to start in building a model neuron, since almost every simplification we might make could lead us into trouble. So, very broadly speaking, a neuron receives inputs from many other neurons at the synapses of its dendritic tree. These inputs are summed, more or less linearly, to produce a single voltage output. One of the first modelling assumptions is that the output should be either zero or one, a Boolean value. In fact real neurons are of two types. There are those with a continuous output, which is essentially a voltage just like a signal in a man-made electrical system. But most of the neurons that make up a mammalian brain are spiking neurons. They work by generating voltage spikes or pulses. These pulses are all roughly the same size, and the rate at which they occur in the spike train indicates the strength of the neuronal output. This is much closer to the Boolean output, but the signal being conveyed is the fraction of time the unit is *on* (spike) rather than *off* (no spike), as opposed to a yes–no vote. This brings in the complication of time dependency.

A real neuron is a physical system evolving continuously in time. Simple neural models frequently are discrete, similar to a cellular automaton where the updates to each element occur synchronously in a stepwise fashion. On the other hand the voltage spikes can occur at any time, not just at specified time intervals. We don't yet know the full importance of asynchronous versus synchronous processing. But we do know that it makes a difference. Cellular automata do not produce the same patterns if the updates are done randomly as opposed to all together. A recent study by May and his colleagues showed that evolutionary systems also look different in these two situations. Even more exciting is large-scale temporal order, which we consider below. Another simplification is to assume that the computation of output from the input is done to high accuracy. Nothing could be further from the truth: real neurons are 'noisy'. Signals are not always totally reliable and may vary slightly in a random way. One might think at first that taking out the noise in a model system couldn't possibly be wrong—the model would just be more accurate. But remember the discussion of Synergetics in Chapter 3. The fluctuations are essential for order to arise.

Our abstract model of a neuron usually has no spatial distribution: it doesn't need one if all that is required is the computation of an output signal based on a set of inputs. Real brains are a spaghetti-like mess of inter-twined cells, essential simply to create the right connections. Now signals which come in from other neurons a long way from the centre of the cell are weakened somewhat on the way in. Furthermore, inter-fering signals (so-called inhibition) can come in on other branches of the cell. The branching patterns are fre-quently fractal and although this could arise just as the result of the growth process it might have some impor-tance in the connection strategy. But we don't know yet if this is fundamental or just an inevitable feature of the biological constraints. Walter Freeman puts this suc-cinctly: a human brain has to live for close to a hundred years and a lot of its construction is concerned with biological maintenance over that time scale.

Learning versus innate knowledge

We have made tremendous progress in the last decade in understanding how learning occurs in neural systems. But artificial systems still differ fundamentally. Mostly they are supervised with error feedback immediately a mistake is made. Some systems do use reinforcement learning, in which feedback is given only occasionally and then in an indirect way. A good example of this is Tesauro's backgammon program. It has to make many moves before it gets just the one piece of data fed back to it: that it has won or lost.

In biological systems, learning takes place at the

synaptic junctions between neurons: precisely how it occurs is still a subject of research, although some proposals such as Hebb's of 1949 still account for quite a lot of neural learning. In the Hebbian model, synapses are strengthened when the neurons they connect fire simultaneously. Thus the neural system builds up a sensitivity to repeated coincidences. But in itself this scheme is inadequate to account for all that goes on: it does not allow for weakening of synapses and it ignores the several different timescales over which learning occurs.

The brain is also full of chemicals, small protein fragments which influence neural communication. Some directly control synaptic activity. Others are more obscure, leading to changes on a more global scale in strange ways. Even nitrous oxide, a reactive, highly poisonous gas, plays numerous roles in the brain. For example, an excess of dopamine is a characteristic of schizophrenia, while a deficiency of it is linked to Parkinson's disease. But although the correlations are well established, neither the mechanism nor even 'which causes what' is certain at this stage. Again we ask whether this sophistication represents an accumulation of tricks during evolution or whether the system would not work, or not work as well if it were simpler.

Training a large homogeneous network to do a very complicated job is difficult. For some simple artificial networks it is very difficult indeed. Judd has shown, for example, that for a class of popular artificial networks it takes exponentially increasing time to get a network to perform bigger and bigger tasks, just as that Travelling Salesman Problem blows out of control as the number of cities increases. That is a real killer, and would indicate that we are never going to be able to build large networks of this kind while they are homogeneous. Thus the need for modular structure, piece-by-piece, task-by-task learning is essential from a computational perspective, mirroring the recent trends in object-oriented programming.

Before leaving the intricate details of neuroscience, we might take in a cautionary tale from a quite different area. Researchers at the University of Sussex have been evolving hardware. They start with a chip referred to as a programmable gate array, in which the connections between its many elements can be user-configured. They use evolutionary programming, which we come to shortly, to evolve the configuration of the chip. It works very well. But the really interesting thing at the time of writing is the effectiveness with which the evolved solutions make use of subtle electrical properties in the chip.

It would not be surprising if it takes us a very long time to unravel all the interactions and synergies in biological brains.

Since Noam Chomsky's famous conjectures on the innateness of language there has been the suspicion that we are born with some predispositions. Chomsky received a great deal of criticism of his proposal but it is now fairly widely accepted. But the human's genetic code is nowhere near large enough to code individual connections' strengths between nerve cells, so this innate knowledge must take some more subtle form. It could be in the large-scale organisation of the brain into different areas. This large-scale organisation does persist from individual to individual in a highly consistent way, so it does seem to be under genetic control to some extent, but local constraints are likely to be involved in some way.

One difficulty with the 'determination of areas' argument is that the cortex is highly malleable. The somatosensory cortex, the area which processes tactile information, exhibits a clear map of the body. But the map is distorted: some parts of the body, such as the hands and fingers, get much more cortex than their simple skin area would suggest. This is not surprising since we have much greater sensitivity and functionality in our fingertips than in the small of the back. Following an accident, say the loss of a finger, some cells will no longer be receiving input. But the cortex now reorganises very rapidly, and the area allocated to that finger rapidly gets taken over by the adjacent fingers. So even if this structure is laid down genetically, it can be very easily changed at any stage of life.

Blood clots within the brain, strokes, cause massive, localised destruction of neurons. But victims, at first badly incapacitated, often make a remarkable recovery, indicating the continuing plasticity of the brain late in life.

In an intriguing experiment, the optic nerve has been redirected during the early life of the subject to some other area of the brain than that which would become the visual cortex. Surprisingly, the visual cortex develops normally but in a different place! So we get the impression that the cortex of the brain is a general purpose processor: what it does depends upon its input and perhaps what is demanded of it. So, returning to Chomsky, perhaps what we have is restrictions on the computing power: only some sorts of operations are possible, and perhaps this is built in at a very low level.

We can get an inkling of what happens from a very interesting idea proposed by Jim Stone. He suggested that the synaptic learning rules are set up in such a way that they are tuned to temporal invariances, things which do not change very much over time. He demonstrated this successfully with a neural network which, without any supervision, learned to recognise stereo disparities. There is a lot still to explore in this exciting area.

Finally, we can ask how the learning rules of a biological neural network themselves evolved. It would seem that there is a driving force to find structure in the world: we are always trying to make sense of the flood of data that comes in through our eyes and ears. But nerve cells are expensive to run. The brain consumes as much as 20 per cent of the body's energy while at rest. Taken together these observations suggest that the neural learning rules might have evolved to develop the simplest and smallest networks to do a job, rather in the manner of finding the algorithmic complexity of an object.

Since we can now make extremely tiny computer chips with awesome performance as number crunchers, data miners, machine controllers, pattern analysers, and many other applications in daily life, we might ask why neural computation evolved as opposed to the digital computers. Of course there is never any answer as to why something evolved. Evolution is contingent and highly uncertain. But Abu-Mostafa at Caltech has an interesting argument as to why neural networks are better than the high speed sequential processors we have sitting on our desks.

He introduces the notion of random problems. In such a problem the rules to solve it are *very* lengthy and complex. Consider distinguishing oaks from ash trees. The shape, size, colour are very irregular. Leaves on one particular tree are similar (to a human!) but that similarity is very difficult to capture mathematically. In such a problem the list of rules and exceptions can be more complicated than just listing all the examples. In other words, these are problems of high algorithmic complexity: they require long programs to solve them. Here, then, is Abu-Mostafa's conjecture: that neural networks are better set to solve problems of high algorithmic complexity than conventional computers are, although they lose out by a long way on speed.

As we learn more and more of the nature of neural computation we will be able to test this idea more thoroughly. It is attractive, though, to think that biological brains may be highly optimised for their role.

EVOLUTION AS A NEW TECHNIQUE IN COMPUTER PROGRAMMING

Just as neural networks have swept through the computing world, so evolution is poised to break into the world of applications after being a minority interest since the 1960s. The area in which it has found immediate interest is the area of computationally hard problems in optimisation (Chapter 2). The principle is simple: encode the problem solution in some way as a genetic code, then evolve the solutions according to some quality level which becomes the fitness of each phenotype. We know that evolution over time has produced some very effective solutions. But is it fast compared to more conventional methods? It is indeed fast up to a level of reasonable approximation, but not when a perfect solution is required.

There are numerous practical difficulties, however. Choosing the population size, methods of selecting the fittest individual and, most importantly, choice of representation all make the method far from guaranteed to work. Nevertheless, the range of successful applications is extensive: job scheduling, timetabling, cellphone frequency allocation and many more.

Finding an optimum

In the days of the first exploration of our planet, intrepid adventurers would hack their way through forests, ford rivers and brave whatever hazards fell in their way, one by one, to the peaks of the mountains. They would eventually get to the top.

Move forward to the early twentieth century and exploration by air. We're looking for the highest mountain peak in a range on which to site a radio beacon. So we drop a hundred paratroopers in a lot of places at random, each equipped with an accurate altimeter. We tell them to climb, and every week we check back to see how they are doing.

Some of our pathfinders have not got anywhere near a mountain and haven't climbed very much at all. We just give them a rendezvous point and tell them to go home. Some have made some progress, though. So now we drop in reinforcements. We don't put them in *exactly* the same places, but put in some little variations, just to try to speed up the search process. Another week goes by and another. Soon we've got nobody left on the plains, everybody is on a mountain. Some might have reached the top of minor peaks and can't go any higher. So they're being overtaken in terms of the altitude reached by others on the higher peaks. So we tell them to go home and we concentrate on the others.

With luck, after a few weeks, we'll have all of them on the higher peaks and soon some will reach the top. With luck, if we started with enough people, we'll have some on the highest mountain and they will eventually outstrip everybody else and find the highest point in the range. In fact there may be several peaks of very similar height. We might not get the absolute best. But we're still likely to find one peak good enough to site our radio beacon on.

This is the principle underlying evolutionary computation. It's a search process; we reward those that do well in the search and eliminate those that do not. The technical details in how we reward and eliminate are a major research frontier in computing at present. These *stochastic* algorithms—algorithms that contain a random component—seem to excel at finding very good solutions to messy problems, which like our mountain ranges have many irregular peaks and troughs.

We talk about a 'search space' for a problem, but what do we mean by it? Searching for a solution makes sense, but where do the peaks and troughs come in? We simply need a measure of how good any possible answer is. In our exploration example it was the height reached. In the Travelling Salesman Problem it was the distance travelled. In finding the optimal allocation of jobs to workers it might be the total idle time added up across all workers. In allocating rooms for lectures it might be how well the assignment of rooms fits into socially acceptable hours (no lectures before 10am, none after 4pm). These are just a few of the endless examples of complicated problems with lots of variables where evolutionary computation works well.

Most people will think of evolution as having something to do with genes or DNA and probably assume that it's linked to sexual reproduction. In evolutionary computation these factors all come in. They are the way we *encode* a problem solution. The genetic code gives us a particular type of eye—for cat, monkey, human—and one day we will understand exactly which genes correspond to what features of an eye. We already know, for example, what genes code for the coloured pigments in the eye's cone photoreceptors, and thus which defective genes lead to colour blindness.

Going back to our paratroopers, let's make the terrain a bit tougher: we'll add snakes, tigers, vicious bugs, steep cliff faces, deep rivers and foul weather. So to survive, everybody would carry guns, ropes, heavy boots, climbing gear, insect repellent—a pretty heavy mixture all up. It's expensive, too, so no single person has all the gear. Of course, the more you carry the slower you progress, but if you're missing a gun when a tiger pops out you might cease to progress at all. So now we add another strategy to our search. Every time we sort everybody into those who have ascended a reasonable distance, we

155

also swap the equipment around. Eventually we end up with someone in the right place (the highest mountain) with the right gear to tackle the hazards *en route.*

In artificial evolution we invent a genetic code specifically for the problem we are trying to solve. The choice of this code is critical and finding good representations is a developing art. This art has been developing over some 30 years, but the theoretical paradigm has lagged somewhat. In fact we are in the rather unusual situation of having an intuitively appealing theorem crumbling under the weight of evidence, with no new theory to take its place.

Building blocks are vital to evolution. The flexibility for complete re-evolution of, say, an eye, is almost non-existent.[13] The vast complexity of living organisms comes from the continual reuse of components: proteins, cells, eyes, ears, limbs and so on. The *schema theorem* of John Holland, the founder of genetic algorithms research, encapsulated this idea of building blocks. Its intuitive appeal does not seem to match its applicability. Although it has not been disproved, the situations in which it works seem to be much more restricted than originally thought.

In the next chapter we shall think about the Internet and its nature as a vast testbed for artificial life. But before we leave evolutionary computation, we'll just briefly mention genetic programming, in which we evolve computer programs. Each gene in our artificial genetic code represents a token in a programming language. The whole genetic code represents a program. Most of the time this will be junk and the program will hardly run at all. Most random programs will be ineffective, but the search process will eventually find some that are functional.

Genetic programming successfully solves quite simple problems and solutions have been reported for more challenging ones. But at the time of writing, scaling up seems to be very difficult. However, we can see the enormous potential for the evolution of programs in cyberspace.

As we have talked of searching, we have implicitly assumed that the landscape is fixed and unchanging. We are looking for an optimal solution to a problem where the value of each solution is known, or can be calculated, in advance. But real biological systems are not like that. Species co-evolve and find environmental niches. Natural disasters take out food, species, habitat in sudden unexpected events. A much richer set of computational properties lie underneath such ecologies. We hope to see them in cyberspace real soon.

The immune system

Nature has another evolutionary scheme besides the one with which we are familiar in the replication of cells. The immune system has to recognise alien invaders in the body, and destroy and remove them. The killer cells generate templates for recognising specific invaders by an evolutionary mechanism, but there is no recombination of genetic material from parents in the traditional fashion. Everything is done by a process of mutation. Another important respect in which immune cells differ from a natural evolution is in their fitness requirements: there must be a huge population, with each cell adapted to some different invader, rather than all cells competing for some common fitness goal. In effect, each cell has its own environmental niche.

Some viruses, such as HIV, influenza and the common cold, are effective because they can rapidly change their protein coat. The immune system is fast and flexible at developing a response to a new invader. But it seems to be limited in plasticity: it mounts a new defence against *old* quarries.

Stephanie Forrest and colleagues at the University of New Mexico have been using the immune system as a model for computer security. The regular plagues of viruses that pass through the PC community cost countless hours of office time, while Internet attacks on enterprise computer systems are a major headache for system administrators. With the rapid growth in the Internet, computer *immunity* is becoming an urgent requirement!

So we have come full circle. From abstract models of computation we build digital computers. But then for some applications we find that simulating biological processes—nerve cell networks, the immune system, evolution—is the best strategy.

CHAPTER 9

THE INTERNET: SUPERHIGHWAY OR GOAT TRACK?

In 1991 US senator Al Gore (later Vice President) made a speech to the US Congress proposing development of a national research and education network. In this speech he referred to the creation of an electronic 'superhighway'. The term caught on. The next few years saw increasing hype in the popular media about the 'information superhighway'. Most of the interest has centred on the potential for video on demand, home shopping and other commercial innovations.

In practice the term 'information superhighway' at present means the Internet. Sceptics have pointed out that the Internet at the moment is more like a goat track than a superhighway. Despite this limitation the Internet is already changing the ways in which we do things in many areas of work and play.

Telecommunications in commerce is not new. Every time you use an automatic bank teller or make an airline booking you are using a computer network. What has excited speculation is the new possibilities created by the development of general-purpose, broadband data networks.

It is worthwhile looking at this phenomenon more closely. The Internet is an excellent example of dealing with complexity in practice. It also provides many examples of emergent properties—one of the hallmarks of complex systems.

WHAT IS THE INTERNET?

So just what is the Internet? Why has it excited so much interest? Behind all the hype we are witnessing an information explosion of historic proportions. The Internet has grown into a vast global communications network. Its origins date back to the 1970s when the US military was looking for ways to improve communication with its contractors. Out of this need arose ARPANET, which soon evolved into a national, and finally international, research network—the Internet.

The Internet is really an emergent phenomenon. Strictly speaking, it is not a network at all but rather a *protocol* for linking together many separate networks. In all there are over 10 000 of these networks, gateways and backbones. Between them they link together over five million individual computers around the world.

With so many computers the number of possible links is enormous—nearly 10^{14} combinations. The system copes with this complexity by organising machines into a hierarchy of *domains*. These are all linked via the Internet Protocol (IP). Every machine on the Internet has an *IP address*. This is a set of numbers that uniquely identify the machine and the various gateways that provide access to it. However, users rarely refer to these numbers directly. Instead they refer to names, which are easier to remember. For example, in the address 'life.csu.edu.au' the domain 'au' refers to Australia, the domain 'edu' refers to educational organisations, the domain 'csu' refers to Charles Sturt University, and 'life' is the name of a particular computer or service. Each domain in the hierarchy has servers that route messages to the appropriate domain or machine.

For most of its early life the Internet existed as a sort of anarchic club. Even among researchers only a few computer-literate souls were able to penetrate its secrets. They banded together to form the Internet Society, which still controls the Internet today.

A transformation came at the end of the 1980s with the invention of several new protocols. Protocols are simply ways of defining how one machine will talk with another. The most widely used protocols reflect the sorts of activities of the users. For computer programmers the most important activities were to communicate with one another (electronic mail), to access other machines (Telnet), and to exchange files (File Transfer Protocol or FTP). One important activity is anonymous FTP: many organisations provide public repositories of free software, images and data. Any Internet user can access this material via FTP, by using the name 'anonymous'. Surprisingly, the US military have maintained some of the biggest public FTP archives.

The importance of anonymous FTP was that it demonstrated the enormous potential of resource sharing. If you make your software freely available, then not only does it save others the time and effort of 'reinventing the wheel' but also they might help by adding to it.

Some examples illustrate this principle in action. The first is FRACTINT, which is one of the world's best fractal programs. Not only is FRACTINT freely available over the Internet but the entire source code is too. As a result, hundreds of people have helped to build and improve the program over a period of years. Recently, Chris Langton and Stephanie Forrest at the Santa Fe Institute have initiated software development projects—SWARM and ECHO respectively—using a similar approach. The SWARM software provides a kit for simulatiing complex systems, using a multi-agent approach. The ECHO project is developing a suite of tools for applying and using genetic algorithms.

However, by far the best examples of collaborative software are the GNU project and the development of

Linux. By the late 1980s the Unix operating system had become the most popular platform for research computing. The GNU free software project started at MIT with the intention of developing quality software for Unix operating systems. One of the most impressive products was the compiler GCC, which turns source code written in the language C into working programs. This program outdid its commercial competitors in almost every respect.

How was this possible? How could a piece of free software be better than the best commercial product? The answer is simply weight of numbers. A large software house might be able to put (say) a hundred programmers to work on a major piece of software. However, this is nothing compared to the forces that the Internet can muster. Try to imagine literally tens of thousands of enthusiastic 'hackers' scattered over the Internet and all vying to outdo one another to improve a piece of code. By the early 1990s the work of this group reached its logical conclusion with the development of Linux. Named after Linus Torvalds, the Finnish scientist who pulled it together, Linux is a complete Unix operating system in the public domain.

As the volume of free information grew, the problem of finding desired items grew ever more difficult. One of the first solutions was Archie, a system of directories named after a comic strip character. It was quickly realised that the process of looking up an entry in the Archie directory and then retrieving the listed item from the appropriate archive could be automated. This innovation pointed the way ahead to developments that triggered the current revolution.

WHAT IS THE WEB?

The need for an efficient way to transfer documents over the Internet is obvious. Many organisations devote a lot of time and effort to delivering information to people. There are huge potential cost savings in having interested people retrieve copies of documents, reports, announcements and so on for themselves.

Several problems plagued early attempts to develop an effective system for transferring documents over the Internet. For a start there was no universal standard for document formatting. Then there was the problem of how to index documents. But mostly there was the problem of developing a simple, automatic way of accessing, retrieving and displaying documents.

The Gopher protocol, developed at the University of Minnesota, offered one solution. In Gopher, the user encountered a series of menus. Each menu item was a link, either to a file or to another menu. The user had simply to click on an item and it would be retrieved automatically.

Gopher demonstrated the power of on-line document retrieval. Also, for the first time, it made on-line information accessible to users who were not computing experts. Almost overnight, a system that was developed to serve as a campus information service became an international information network.

Unfortunately, Gopher suffered from three major limitations. Firstly, its menu system was rigid and at times frustrating. Secondly, it did not prescribe a formatting standard for documents. Thirdly, indexes and documents were separated.

Meanwhile, researchers at CERN in Switzerland developed a different approach, the HyperText Transfer Protocol (HTTP), which overcame the above difficulties. In hypertext, words or phrases in a document can be made into 'hot links' to other documents. This means that we can incorporate Gopher-style menu items directly into a document. Moreover, HTTP prescribed a clear formatting standard as well—the HyperText Markup

Language (HTML). In computing terms, it operates via a client–server model. That is, a user's program (the *client*) sends a request to an HTTP *server*. That server then sends back the requested document to the client program, which displays it for the user.

The crucial feature of HTTP is that it provides links between on-line documents. Sets of documents that are linked in this way can be thought of as 'clusters'. As we saw in earlier chapters, the formation of linkages and clusters is at the heart of self-organisation in complex systems. So in a very real sense HTTP provides a mechanism that promotes self-organisation of information on the Internet.

Having developed HTTP, the CERN group set up an experimental set of sites for their system, which they called the World Wide Web. For a time the Web seemed doomed to oblivion for want of an effective interface. Then, in 1992, the US National Centre for Supercomputer Applications (NCSA) developed a brilliant program called MOSAIC that made the Web come alive. The number of users and publishing sites began to grow exponentially, and the rate increased markedly when browsers for PCs and Macintosh were developed. At one stage the number of users was doubling every nine weeks. By the mid-1990s Web activity had become the dominant traffic on the Internet. The World Wide Web is now firmly established as the Internet's publishing medium.

Many hands make light work

The development of information services on the Internet exhibits both self-organisation and emergent properties. We can best appreciate this from some examples.

The Internet is a complex system that is just beginning to organise itself. One organising trend arises from the need to be able to draw together information from many different sources about a particular subject or for a particular purpose. Many people and organisations compile databases of references about their area of interest. In some instances this activity can create new resources where the whole is quite literally more than the sum of its parts.

A good example is the Virtual Tourist. This information system allows users to zoom in from a map of the world to any country and call up basic information. Very quickly the user can be checking pictures of (say) the Botanic Gardens in Edinburgh or the current weather in Toronto. The point is that the Virtual Tourist is not managed by a single agency. It arose by systematically linking together information services from literally thousands of different Web sites all over the world.

Another emergent service on the Web is the Virtual Library. This is really a huge table of contents for the Web. Each topic in the directory, such as 'complexity systems' or 'computing', sends the user off to a server that maintains a comprehensive index of links about the subject concerned. Like the Virtual Tourist, the service integrates information from tens of thousands of separate Web sites.

The Internet is also influencing people's activities in areas away from the computer. The achievements of researchers in molecular biology provide an excellent example. In the late 1980s biologists set up several major databases on the Internet as repositories for biological data, especially gene and protein sequences. Prominent examples include Genbank in the United States and EMBL in Europe. It is now compulsory for researchers to submit their raw data to public databases such as these if they want to publish their research in any leading journal.

The present situation has arisen for a good reason. By adding together everybody's data we get more than we started with. Not only can scientists use the databases to help them interpret their own results but they can carry out entirely new kinds of study. For instance, they can search for patterns in the vast pool of data about DNA and proteins now available within these databases.

The above examples suggest ways in which information services on the Internet are likely to develop in the future. In most fields of activity there are enormous potential benefits in developing joint resources of the kinds described above. As a result we are already seeing many attempts to create 'information networks', 'virtual communities' and other cooperative projects. It is easy to identify many issues, such as documenting the world's biodiversity and studying global climate change, that call for joint efforts of the kind we already see in biotechnology.

The knowledge web

The Internet is perhaps the ultimate complex computer. At present, the bulk of traffic across the information superhighway consists of documents, images and electronic mail messages. However, linking computers together makes it possible to spread the burden of calculations across several machines. The Internet is ideally suited to distributed computing, where different machines specialise in solving different aspects of a problem. We are already seeing this phenomenon with the World Wide Web. What happens is that organisations publish information about their specialised area of knowledge. For example, the Weather Bureau publishes daily temperatures, the stockmarket publishes share prices, and so on. This means that we can draw together reliable, up-to-date information about a subject by consulting services provided by organisations that deal with each aspect of the problem. Suppose, for instance, that you want to compile a report about an endangered plant. You can obtain basic taxonomic information from the botanic gardens. National environmental agencies provide data about its distribution and about conservation measures. Universities and government research organisations provide details of current research about the plant.

The above sort of specialisation is now spreading to include many kinds of computation. For instance, in the course of their work scientists, engineers and other professionals often put together computer programs to carry out sophisticated and specialised tasks. In many cases the software would be of use to many other people, but the development of such programs is extremely costly and time-consuming. However, non-specialists may only need to use the particular calculation involved once. Therefore it is not worth their while buying an expensive piece of software for the purpose. As an alternative, organisations can now provide a user interface to their specialised software and services on the Internet.

In the above scenario, Internet computing is no different from what already happens in many industries where a project manager calls in a host of specialist contractors. However, by automating and standardising the process we are likely to see Internet computing go much further. Suppose that a query system does not need a user to sit down and fill in details of the processing required. Suppose instead that it accepts the job as a bundled query 'object' from another computer, and that it sends back a result 'object' in the same way. Then we could write programs that include automatic queries to other machines.

All of the above possibilities are already being explored. However, if the theory is straightforward, the practice is not. At present, the Internet is a disorganised mess. While it is true that we can now combine information from many different sources, locating that

167

information is a huge problem. At the moment, the most common way of locating information is via one of the main Internet search engines. These are huge databases that list the contents of thousands of sites. Sometimes this can be a useful way of discovering previously unknown resources. However, it can also be extremely frustrating. For example, there are tens of thousands of references to Shakespeare on the Web; and if you want to look up information about flu and other bodily viruses then you are likely to be swamped with references to computer viruses.

We have alluded to two processes that are promoting organisation. These act as filters, in which similar information resources are linked together. Likewise, there is the need to be able to draw together information from many different sources about a particular topic or for a particular purpose. So people and organisations are compiling databases of references about their field of interest.

One of the driving forces behind the present anarchy on the World Wide Web is the increasing commercial activity in which companies see the Web as a means, not for sharing information, but for promotion and advertising. So besides promoting an intense aversion to coordinating and sharing on the part of information providers, it has led to an incredible duplication of effort. The reason is that the quickest way to promote your Web site is to provide material that will attract users, even if that material is already provided elsewhere.

For instance, one of the services that the authors provide on the Internet is a 'Guide to Australia'. This is a distilled index to essential information resources about Australia. However, over the past two years almost every service provider that has set up on the Internet has duplicated this service in one form or another. Despite the current emphasis on sites, rather than services, there is great pressure from many directions to

168

organise information on the Web. One of these pressures comes from the need for organisations to coordinate their activities. Libraries, for instance, need to share information about their holdings—for handling interlibrary loans and so forth. Similarly, many government and scientific organisations need to share essential data and resources. This has led to the spontaneous formation of many professional information networks, virtual communities and other similar mechanisms for coordinating the provision of information. In time, hopefully, the present preoccupation with individual sites and 'home pages' will decline and topics and activities will become the focus of organisation. In the coming century the above processes should see the Internet grow into a comprehensive, worldwide information system. In the scientific arena we can see a clear historical progression in the way discoveries are communicated. In the time of Newton and Galileo letters were enough to keep colleagues informed of new results. By the nineteenth century we had the age of the great scientific societies, with discoveries announced in formal lectures to the Royal Society. The twentieth century saw the heyday of the scientific journal. The twenty-first century will surely be the age of what we might describe as the 'knowledge web'.

All of the above sounds absolutely thrilling and there is reason to expect that the Internet will improve our lives and make many new activities possible. However, there is also a darker side to the current information explosion. On the Internet we are already beginning to see the emergence of what can only be described as 'information imperialism'. First of all, the Internet uses English as its standard language. This is reinforcing the position of English as the prime language for international communication. It is a harsh reality that anyone whose Internet site provides a service in any language other than English will simply not attract a large international audience. Secondly, the Internet is serving to promote American ideas and culture. This is a function of the vastly greater number of sites and services emanating from the United States than anywhere else.

Finally, and perhaps of greatest concern, the Internet is tending to deny people a voice. This may sound a bit surprising at first. After all, on the Internet everyone is connected to everyone else. It is true that the Internet can level the playing field in some respects. Even a 'backyard' server has as loud a voice as the largest corporation. However, the problem is really whether people will look at what you have to offer.

Here we have the problem that network speeds are generally much faster in the United States than in many other parts of the world. Hence users expect very fast response times. They become impatient when services are slow to respond. This raises a problem for sites outside the US because international links are generally heavily used and generally slower than local links. So even though many countries have very comprehensive information and up-to-date and easy-to-use services about themselves, American users prefer to get the story from an American source. This situation is compounded by indexing sites pointing to those US sites, rather than the more authoritative and up-to-date sites overseas.

This aspect of the Internet reflects a similar process that is also under way in many other aspects of modern life, wherein the diversity of business and cultural resources is being overtaken by monolithic international corporations.

CYBERSPACE

You do not have to be a good scientist to write good science fiction. In a fast-changing field—and computers are probably the *fastest*-changing field at the moment—

knowing too many of the details can cripple the imagination. William Gibson, inventor of the term *cyberspace*, was not short on creativity. His characters inhabit twilight worlds of physical reality and an information space, in fact something like the Internet. But whereas Web servers are static and unintelligent, in cyberspace there is not only information but also intelligent beings.

Gibson's cyberpunks connect to the 'Internet' through direct brain implants. Farfetched? At present, maybe, but not outside the realm of imagination. We know, for example, that the cortex of the brain is at birth just a general purpose neural computer, which can develop in lots of ways. As noted earlier, one of the more intriguing (and maybe rather gruesome) experiments of recent years has been the demonstration that one can mix up cortical development.

Suppose that you surgically remove the optic nerve and plant it somewhere else in the cortex, say in the area destined to become the hearing centre. The amazing thing is that that part of the cortex now develops as a perfectly functioning *visual* area. In other words, what the particular piece of cortex does depends upon the input it receives. Now, the cortex has evolved expecting a visual input *somewhere*. One of the big questions in neuroscience at the moment is in what way evolution has prepared the brain to expect particular sorts of input. One definite possibility is that the main predisposition is towards recognising *structured* information—that it doesn't really matter what the structure is (whether it's visual, auditory, tactile or any other sort of information) as long as it contains patterns.

In this eventuality, the way is open for *cyberdecks*, where the brain plugs directly into cyberspace. It seems unlikely, though, that this would ever work for an adult brain. Thus the ethical conflict for experiments of this kind would never be resolved. The operations would have to be carried out at a very early age, before the child had any vote as to whether he or she wished to be a cyberpunk!

Intelligent agents

So how many years are we away from the cyberpunk era? Perhaps not as many as we might think. Let's track the initial stages of the evolution of Wintermute,[14] the giant Artificial Intelligence in *Neuromancer*, one of the first Gibson cyberspace novels. The evolutionary force to begin with is the enormous but chaotic growth of the web. Lots of people have trouble reading all the email they are sent. Push technologies, which dump more and more information on the desktop of your computer, are already appearing. So we need to build filters, clever filters which know what we like and don't like.

Companies such as Firefly in the United States already build profiling agents. We can do a lot with simple statistical matching. For example, word frequencies are amazingly discriminating in characterising author style, so much so that experts are willing to stake money on whether Shakespeare did or did not write particular plays and so on. It just turns out that we tend to use some words more often than others (I'm particularly bad with 'particular'). So we could make a list of keywords that describe our interests and then look for frequencies of these words in email messages or web pages. This is what we might call a strong AI approach. We're defining an algorithm (a statistical formula and a data set) to do the job.

The above agents can be quite successful. But we can expect different things down the track. One alternative way to build an agent is to make its parameters *evolve* : so we might have a formula which determines how we do a match between the things we write on our own Web site and the things on other sites. Our agents could evolve this formula to make it more and more useful.

It would work something like this. When we do an Internet search, for example, we would be given an agent (that's one example of the formula) which would go away and do the search. Then when it returns we would give it a ranking, say by clicking on a score. A daemon sitting in the background would continually be mating agents against one another according to their scores, in a 'survival of the fittest' setting.

But we saw earlier that genetic programming is a reality: we can actually evolve computer programs out of random strings! So why not go one step further? Instead of specifying a formula at all, simply evolve the method the agent uses from scratch. We might need a lot of patience at the beginning, but it could pay off in that we might find solutions we never dreamt of. Although this seems to be suggesting that evolution can find better algorithms than we can create ourselves, this does in fact happen. Evolutionary researchers frequently see their programs finding unexpected solutions, often because they overlooked something in their experimental design—just like the undoing of Clever Hans.[15]

So far we are still in the present. All these sorts of ideas can and are being implemented here and now. But how do our agents get real autonomy? The secret has to lie in making the criteria for success, the fitness criteria, more and more general. It's like the difference between a secretary from the temp. staff agency and a personal assistant who has been with you for years. The instructions are fewer, the feedback is on a much longer time scale and there's a vast difference in the usefulness of the two. But in allowing the fitness criteria to become very general we are allowing our agents to evolve in perhaps unexpected ways.

The final stage is defining protocols on the Internet which enable agents to query one another. These automatic agents extend the idea of query objects, which we discussed earlier. We already have so-called 'common object busses' and object request brokers which enable computing objects to talk to one another, even though they may be on different machines of different architectures and be written in completely different programming languages.

Remember that the idea of an object is a chunk of data which is protected from the outside world and accessible only through special procedures, or methods, that belong to the object. In response to a message from outside, a method will return a piece of data, perhaps customised in some way.

The stage is set for the evolution of communication and independent action of agents and objects on the Internet. They can become a self-perpetuating computational ecosystem. From that point on, we lose control. Our only option would be destruction, but that is not likely to happen. The growth of intelligent agents will be fuelled simply by their usefulness. We will be reluctant to let them go.

CHAPTER 10

COMPLEXITY AND LIFE

What will industry be like a hundred years from now? It is sobering to try to guess how science and technology will develop in the future. Almost every area of technology has advanced dramatically during the twentieth century. It is reasonable to assume that progress will continue. However, if the past is any guide, much of that progress will be the replacement of one technology by another.

We have also seen the rise of entirely new technologies. At the turn of the nineteenth to the twentieth century, motorised transport gradually replaced horse-drawn vehicles. The twentieth century has seen the rise of electronics, computing and aeronautics, to name but a few new industries. At the century's close many more technologies lie on the horizon. For example, although it is already very active, biotechnology is poised to spawn major industries during the twenty-first century. Likewise, space travel, and a host of possible technologies that exploit micro-gravity, are still in infancy. In the context of this book, the question of greatest interest is what new discoveries and what new technologies might stem from the development of complexity theory. As we have seen, discoveries about the nature of complexity in general have already led to new insights about specific systems and processes. Studies of chaos are helping scientists to understand the nature of turbulence. No doubt there are many more discoveries to come.

When we consider technology, complexity has a lot to offer. Perhaps the greatest prospect is the possibility of learning how to control processes that involve really complex interactions. Take biotechnology, for instance. The essence of biotechnology is to duplicate organic processes in an artificial environment. The great breakthrough offered by genetic recombination is that it is now possible to manipulate and adapt natural processes to our own ends. Impressive as they are, the current generation of applications are really fairly simple. A good example is the production of interferon, a substance that is produced naturally by the body in small quantities. It plays an important role in helping the body to ward off disease. Almost as soon as it was discovered, the potential value of interferon in medicine was realised. The problem was the extreme difficulty of extracting enough for medical use. Genetic recombination solved the problem. By splicing the fragment of human DNA that coded for interferon into bacteria, the bacteria were persuaded to manufacture interferon as a by-product of their metabolism.

The above one-to-one correspondence of spliced genes to single products can only take biotechnology so far. A much greater challenge is to learn to emulate genetic control over development. Most stages of development and growth require ordered interactions between many genes. The work of Kauffman and others is beginning to reveal how growth processes are organised at the genetic level. The ultimate possibilities that a full understanding of growth would lead to are simply breathtaking. Many machines will contain organic components. Many materials, and even complete products, could be grown instead of manufactured.

UTOPIA OR BRAVE NEW WORLD?

In the above discussion we have looked at just some of the possibilities opened up by new technology in the coming century. However, we should stress that our discussion is definitely not meant to be an argument for blindly embracing new technology. Recent developments in medical technology—cloning, embryo transplants and the like—have brought into sharp focus the question of whether or not we should adopt new technologies merely because they are possible. Any new technology has two sides. On the one hand it solves problems or creates new opportunities. On the other hand it also has an impact on the *status quo*. Its effects can be economic and social, as well as technical. Moreover, these effects may be negative. So any benefits that a new technology conveys should be weighed up against the costs. New technologies have often had undesirable side effects. Let's look briefly at some examples.

The invention of television was one of the great achievements in communications during the twentieth century. Although TV has provided great benefits, it has also had a huge impact on society. In the Western world it is arguably a major factor in alienation and social breakdown. In Third World countries it has promoted a widespread partiality to Western culture, disrupting local customs and tradition. Improvements in transport have a similarly chequered history. In Western cities fast, cheap public transport has led to the development of vast suburban dormitories. The average standard of living has increased. However, like TV, transport has also contributed to social breakdown. People move further and more frequently. Friends and family often live far away. People have less attachment to their local community and fewer social contacts.

177

In industry, automation, such as the introduction of robots, has increased production and helped to greatly reduce the costs of products to consumers. But these advantages have also contributed to enormous cuts in employment in the industries concerned. Perhaps the greatest advance of all has been the enormous improvement in public health. In most parts of the world, life expectancy has increased and the traumas of disease and death in childhood have been greatly reduced. Great though these benefits undoubtedly are, they also have a negative effect: overpopulation. For almost a century now the world has been suffering from runaway population growth. The results are widespread poverty and suffering, plus periodic incidents of famine and starvation.

Some of the worst stories arise from mismanagement of the environment. We simply do not know enough to get it right. Earlier we saw the example of the cane toad in North Queensland. Environmental management is full of stories of misguided schemes that went wrong. Genetic engineering raises the spectre of even greater environmental disasters. The fear is that genetically altering (say) a virus to control some pest might produce a supervirus that attacks other organisms, and even humans.

The recent story of the calicivirus illustrates just how potentially disastrous genetic manipulation could be. Rabbits have been a major environmental problem in Australia for over a century. For many years the disease myxomatosis kept numbers in check. But by the 1990s the rabbits had become resistant and their numbers began to increase again. To replace it the CSIRO[16] genetically altered a calicivirus. Aware of the dangers, they carried out an experimental release into a rabbit population on an island off the southern coast. Almost at once the virus mysteriously started turning up in rabbits on the mainland. No one knows how it got there, though desperate farmers were suspected. Within weeks the virus had spread over a vast area of the mainland rabbit population. Fortunately, there seem to be no obvious side effects. However, the risk is that one day such a virus might cross over into other species and wipe them out too.

THE ANATOMY OF AN ACCIDENT

It is a dark and stormy night. Rain is teeming down. At a busy highway intersection a driver who is hurrying home starts to make a turn. Out of nowhere another car suddenly appears. It is coming straight for him. Both drivers slam on their brakes, but cannot stop in the wet conditions. There is a sickening crunch and both drivers end up in hospital with serious injuries.

Why did the accident occur? Whose fault was it? Such questions are often hotly disputed. They can be the subject of much passion, anguish and litigation.

The truth is that accidents are complex events. They often result from a deadly interaction of many factors. One popular saying is that accidents happen when things are different for one reason or another. In the above example, the storm is a major difference from normal driving conditions. So is the pressure on at least one driver to get home quickly.

In general, accidents occur through the interaction of different factors. For example, everyone has a *reaction time* —the time it takes to react to a new situation. Potentially dangerous situations, such as cars pulling out into our path, occur all the time. Normally our reaction time is perfectly adequate to cope—otherwise driving would be impossible. However, many factors can affect our reaction time. For instance, tailgating, where one car is moving along very close to the one in front, can reduce the available time to react to virtually zero. So can poor light and wet roads. Also other factors, such as distraction or fatigue, can slow down the driver's reactions.

Any combination of factors makes for a very dangerous situation indeed. We may be able to cope with (say) wet conditions most of the time. But in extreme situations, where we need to perform best, we are most likely to fail. And these extreme conditions are more likely to occur when other factors, such as bad roads or driver fatigue, enter the equation as well.

Deadly combinations of factors are well known to air crash investigators. Sometimes a chain of events builds up, with fatal results. For example, the crash of an aircraft into the Potomac River, Washington DC, in the mid-1990s involved delays in takeoff, due to snow and ice. This put pressure on the flight crew. After first forgetting to turn on the de-icing equipment, they rushed through their checklist and failed to act on the error. Instead, they tried to take off anyway.

Accidents in the home often build up in a similarly tragic way. In a mad rush to get off to a party, Mum and Dad leave electric hairdryers and razors connected and lying on the bathroom bench. The babysitter, being a stranger, assumes that all is safe when she sends junior off to the bath. But the toddler, finding a new toy, plunges the hairdryer into the bath and is promptly electrocuted.

The events leading up to an accident can be long and complex. They can involve the interaction of many small events and circumstances, which by themselves are not significant. People like things to be simple. We always try to identify simple patterns of cause and effect. So the complex nature of accidents can be both confusing and frustrating.

In trying to understand the nature of cause and effect, philosophers used to talk about 'immediate' and 'final' causes. Thus, in the above example, plunging the hairdryer into the bath is an immediate cause of the accident and leaving the hairdryer turned on is a final cause. However, as we have already seen, there can be many contributing factors, so even this categorisation is too limiting.

None of this is helpful to the relatives of victims, as they try to come to terms with their loss. There is a tendency to pin blame, to seize on one of the many factors that may have been involved and present it as 'the cause'. It is all too easy to put the blame on a dead pilot, for instance.

Lawyers, with an eye on larger fees, are all too easily tempted to place the blame for an accident wherever there is a potential for a payout to their client. In the United States, this habit has led to some unfortunate and bizarre results.

One result is a tendency to sue doctors the moment anything goes wrong. The factors contributing to illness and death can be just as convoluted as accidents. The idea of malpractice is to guard against gross negligence, such as sawing off the wrong leg. But where does negligence end? In the extreme, one can argue that there is always something more that can be done to save a dying patient. So it can be argued that any time a patient dies the doctors are *ipso facto* negligent. Huge payouts in malpractice suits seem to be commonplace, especially in the US. The fear of litigation affects the way in which doctors work. For instance, obstetricians who are anxious to avoid complications have become increasingly interventionist. One sign of this is an increase in the frequency of Caesarean sections.

The risk of litigation also leads to a feedback effect in which higher payouts force higher insurance fees for doctors. Doctors have to pass on those costs in the form of higher fees for service and their patients therefore have to pay much more for medical insurance. Poorer families cannot afford the rates, and avoid visits to the doctor. Their standard of health goes down and death rates go up.

Litigation related to accidents has reached bizarre levels. If a burglar trips over a rake in your backyard, he can sue for damages. But more worrying still is a disturbing trend that emerges from accident payouts. Just as with medical malpractice, individuals and companies are liable if gross negligence leads to injury or death. But again, what is negligence? Is it negligence if a customer rushing out of a store slips on paving wet from the rain and injures her back? Should the company have anticipated that she would have been rushing and put up signs urging caution?

How far should this sort of thing go? What sorts of dangers can an organisation reasonably be expected to anticipate? If a car crashes through the front of a shop, is the shopkeeper liable for damages to customers because he or she failed to install safety glass in the shop window? If a diner at a restaurant gets sick from overeating, is the restaurant at fault because it provided the food? If a man injures his back in a novelty contest where entrants have to run a race carrying a refrigerator on their backs, is the manufacturer at fault because it failed to warn the man that such activity could be dangerous?

What we are seeing here is a gradual process in which doctors, airlines, manufacturers—in fact everyone providing any sort of service at all—are expected to assume more and more responsibility for the health and safety of the population. Conversely, the sorts of cases we have seen above contribute to a culture in which individuals are encouraged to assume no responsibility whatever for their actions.

All of this contributes to an increasingly consumerist culture in which individuals expect everything, including care, to be provided for them. Consumers assume no responsibility at all. One could reasonably argue that it is a civic duty to alert fellow citizens to dangers. If a shopper falls over on wet pavement, why not sue the passers-by who failed to alert her about the slippery pavement?

In hindsight, almost any accident is avoidable. The underlying problem is one of complexity. There are always so many combinations of factors that we can never anticipate them all. Also, some dangers can be anticipated but are so rare that to try to counteract them all would be prohibitive in time and effort. For instance, on any suburban street there is always a chance that somewhere, sometime, a child will dash out from behind a parked car and be hit by a passing motorist. This does happen. Should every street carry large warnings? Should we restrict speed limits to 5 km/h to ensure that drivers can react?

This brings us to the general issue of prevention and precaution. Some dangers are obvious and real. Building codes include measures to reduce the risk of fire. Likewise there are codes for occupational health and safety, for food preparation, and for a hundred and one other things. But how many precautions can we take? To take an extreme case, what should we do about the danger of meteorites? People have been killed by falling meteorites. Anytime, anywhere, there is a finite danger that a meteorite will strike. It may be small, but it is real.

One issue that modern culture has not dealt with very well is balancing risk with reaction. We've all heard stories about officials failing to install crossings or take some other action on safety until someone has been killed. Conversely, there are many examples of over-reaction. Air hijacks are extremely rare events, compared to the total amount of air travel. Nevertheless, they attract such vast media attention that airports everywhere now spend millions of dollars every year on security.

183

Dealing with accidents has played a surprisingly large role in shaping society. Modern society has many institutions that deal with accidents in one way or another. Examples of mechanisms to prevent accidents include traffic laws, building codes, regulations for food preparation and for occupational health and safety, and drug testing. And there are the institutions that respond to accidents, including ambulance, fire brigade, search and rescue units, and police.

Just as the increase in complexity of our society brings with it increased risk of accidents, so the change in managerialist culture brings with it hidden dangers. Biological systems are highly adaptive, but they also possess considerable *redundancy*. Engineers have understood this for a long time. You can't expect a message to be transmitted across a communication channel perfectly all of the time. So what you do is to build in extra information contrived in such a way that if any particular bits are lost, it is still possible to work out what they were. In the times of massive over-employment we had considerable safety checks built in to organisations simply because there were enough people around with enough shared knowledge to catch other people's mistakes before they became life-threatening. The move to ever greater efficiency has cut out much of this safety net. A tragic example was the *Herald of Enterprise* disaster, in which a sea-going ferry sank, drowning dozens of passengers. This occurred because one person had ultimate control of the ferry doors and, one night, he accidentally left them open. If we have to make organisations more efficient, we also have to redesign the safety nets. *Until we fully understand the dynamics of complex systems we may not be able to do this properly.*

In the above stories we have tried to provide examples that show how *complex* phenomena such as accidents

184

influence society. The immediate effects can be subtle, and perhaps minor, and yet the end results can be profound changes in our social fabric.

SYNCHRONISING SOCIETY

How many people do you know? For most people the answer would run to scores of friends, family, colleagues and acquaintances. If you count everyone you ever met, then the answer could easily run to hundreds of people. However, if we restrict the count to individuals in regular (if infrequent) contact, then the number rarely goes beyond 200.

In Chapter 6 we saw how interactions between people define connections and networks within society. Just how influential are networks in shaping human culture and society? To look at just one type of network, we can ask what role gossip plays in human society. The instant response is almost invariably to assume that gossip is a trivial activity. However, Robin Dunbar, Professor of Psychology at Liverpool, suggests that gossip has played a crucial role, not just in the history of human society but also in the evolution of language and of humanity itself.

Gossip is the everyday exchange of ideas and pleasantries. We associate it especially with the spread of stories about social matters. Dunbar suggests that gossip evolved in early humans as a kind of verbal grooming. Now, it is well known that grooming in ape societies is an important social activity. It helps to maintain social cohesion. Dunbar's theory is that, at some point in human evolution, speech began to take over the role that grooming plays in troops of apes and monkeys. He goes on to suggest that this new role was an important factor in the development of speech and hence in the evolution of *Homo sapiens* as a species.

As a form of grooming, speech has the advantage that you can talk to several people at once. The one-on-one grooming seen in (say) baboons or chimps is slow. To interact with every member of your troop by grooming takes a long time because you have to groom them one at a time. However, if you can talk, then you can interact with up to three or four other individuals at one time.

Suppose that you're talking socially with a few friends at a party. Then if the group is small (less than five), everyone joins in the conversation. But if a couple more people join the group, it quickly breaks up into two separate groups, each discussing different topics. This sort of phenomenon suggests that there is a natural maximum group size (about four) for social discussion.

Now if we suppose that gossip is social grooming then we should find that the size of human clans is larger than ape and monkey troops. It is well documented that apes and monkeys form troops of about 30–50 animals. Dunbar has gathered a lot of evidence to suggest that there is a natural troop size for humans of about 100–150 people. He cites examples such as the Roman *century* and the fragmentation of religions into communities as cases in point.

This entertaining concept has a neat physiological twist to it. Why should grooming bind animals together? It transpires that grooming not only serves a useful function, in, say, removing twigs and burrs from the fur, but it also does something unexpected. It causes a release of the natural opiates in the brain, creating a pleasurable experience. It appears that laughter also releases natural opiates, thus allowing language to bind animals together in the same way as physical grooming, but on a larger scale.

Gossip does play an important role in shaping community attitudes, especially (say) in small towns or other groupings. Conversely, the same theory would argue that lack of communication is a major contributor to the problems of alienation and antisocial behaviour that we see in many Western cities. The erosion of traditional religions and the rise of TV culture have presumably contributed to this process.

As far back as we can trace, people seem to have had religious beliefs. These beliefs have exerted an incredibly powerful hold on individuals. Saints have suffered uncountable horrors, Abraham would have sacrificed his son, while hundreds engaged in mass suicide in the Jonestown religious cult. Do we learn anything from complexity about how religion can produce coherent behaviour among human beings? We might suggest, tongue in cheek, that Prophets embody the slaving principle (see Chapter 3). They convey a message so powerful that others follow regardless of their own desires. Other religions have rituals, varying in frequency and depth, which provide the synchronisation keeping groups together.

But it is internal codes of action, be they religious or otherwise, that can produce distinctive societies. Changing the rules and extent of interaction will produce just as strong a social pattern as rigorous central control. But as with the other complex systems we have seen throughout the book, working back from desired pattern to rules is far from trivial, if not impossible.

LIFE THROUGHOUT THE UNIVERSE

The year was 1976. After a journey of over 150 million miles the first Viking probe was about to land on Mars. Interest was intense. For the first time ever a probe had been sent to another world with the specific aim of searching for signs of life. Both scientists and an interested public eagerly awaited the first photos. What would they show? Would we see actual living creatures from outer space? Would science fiction at long last become science fact?

The history of interest in life in outer space goes back a long way. Ever since astronomers first became aware that the planets were really other worlds like the Earth, people have speculated about whether or not life (and particularly intelligent life) exists elsewhere in the universe. The expectations of finding life on Mars took a severe battering in 1965 when the Mariner 4 space probe swung past Mars. The first close-up photos of the planet showed a barren surface pockmarked with craters. Mars suddenly seemed more like the Moon than the Earth. Instead of little green men, microbes seemed to offer the best hope of finding life on the red planet.

This, then, was the background to the Viking landings in 1976. When the first of two landers touched down on the surface, no one really expected Martians to come marching up to the machine. There was still a real hope of seeing lichens and other primitive plants. But there was little surprise when the first photos showed a barren, rock-strewn landscape. At least the sky was pink.

Much more faith rested with the lander's shovel, which scooped up a sample of Martian soil for analysis in a robot laboratory. The scientists planning the Viking mission had devised what they thought would be a series of definitive experiments to test conclusively whether organisms lurked in the Martian soil. These experiments included tests for three of the main characteristics of biological activity (on Earth)—respiration, growth and metabolism. In what is perhaps the best tradition of science the results confounded everyone. Two of the experiments gave negative results, but one was positive. No one had anticipated anything like this. Something very strange was happening in the Martian soil. The

Viking experiments prompted a lot of fresh thinking about what was going on. One theory was that perhaps the Martian soil was like soil on the early Earth. On Earth today, air, water and soil are almost everywhere affected by the action of living things. Perhaps on Mars there were certain chemical reactions that on Earth were once quite common but have been obliterated by the activity of living organisms. All the same, the essentially negative findings of the Viking mission all but ruled out Mars as a serious candidate for hosting life. Attention therefore shifted to other possibilities, both within the solar system and beyond. Associated with this shift was an increase in interest in the origins of life.

For a long time the abundance of carbon dioxide and nitrogen in the atmospheres of Mars and Venus was taken as evidence of their similarity to the Earth. The assumption was that in the early solar system all three planets had atmospheres that resembled those of the outer planets. The outer planets have an abundance of methane and ammonia in their atmospheres. The theory was that the activity of living things had transformed the atmospheres of the inner planets to give them their current composition.

In 1953 Harold Urey at Chicago University performed an elegant experiment to test this theory. Into a set of connected flasks he placed water in an atmosphere of ammonia and methane. He then passed electric sparks through the gases of the flask to simulate the effect that lightning might have had in the early atmosphere. After a week the water had turned into a pink soup laced with a wide variety of organic compounds. These compounds included amino acids, the building blocks of life.

These results provided a perfectly consistent scenario for the origin of life on Earth. Over millions of years lightning storms in the atmosphere would have produced vast reservoirs of organic matter, creating an environment apparently ripe for the appearance of life. Now that probes have carried out close encounters with most of the planets, it is apparent that differences in their composition and atmospheres are linked to the origins of the solar system. In the condensing clouds of gas and dust that formed the early solar system the inner planets accumulated a very high proportion of heavier elements, whereas the outer planets acquired high proportions of gases. As a result the original atmospheres of Venus, Earth and Mars were quite different from those of Jupiter, Saturn, Uranus and Neptune. Current thinking is that the atmospheres of Venus and Mars today retain essentially the same composition as they had when they formed. The Earth, however, started with an atmosphere very similar to those of the other two, but living processes have transformed it to give the present composition. It is now clear that the clouds of Venus are the end result of a runaway greenhouse effect—and a dramatic example of positive feedback in action. Venus receives nearly twice as much radiation from the sun as the Earth does, but the dense clouds reflect much of it back into space. That is why Venus is the brightest 'star' in the night sky. However, the telling factor is that the dense atmosphere and heavy cloud cover retain heat, allowing the surface temperature to climb to a hellish 800 degrees Celsius.

There is an important lesson in this. Here is a planet that is virtually the twin of the Earth and there is every reason to suspect that Venus might originally have had all the ingredients that led to the appearance of life on Earth. And yet the Earth is ideal for life, whereas Venus would fry any living thing in seconds. Perhaps this is a cosmic lesson that we should heed. It teaches us that the greenhouse effect is not a figment of some alarmist imagination. Unless we act there is a very real risk that global warming arising from the build-up of greenhouse gases in the Earth's atmosphere will become irreversible. In the worst case scenario the Earth could literally end up like Venus and life, all life, on this planet would end!

LIFE AND ITS ORIGINS

To understand the origin of life we first have to consider the underlying chemistry. It begins with mechanisms that *enhance*, not suppress, order and pattern. Ilya Prigogine was concerned with the way in which self-organisation occurs in physical systems. He focused particularly on *dissipative systems*. These are systems that produce energy internally and exchange it with an external environment. They include living organisms and many chemical reactions.

Prigogine pointed out that, in dissipative systems, minor irregularities can grow into large-scale patterns. That is, such systems are highly sensitive to random variations. This means that in a prebiotic world dissipative systems would have provided ideal conditions in which novel chemical structures could appear, spread and diversify. An environment that suppressed chemical variation would have prevented complex organic compounds from appearing at all.

Organisms must have a source of energy. On Earth the two main processes are photosynthesis, in which plants trap light energy from the sun, and respiration, in which cells 'burn' carbohydrates and fats. The assumption has generally been that life elsewhere in the universe must use the same processes as we see here. However, the discovery of organisms that metabolise sulphur and rely on the heat from volcanic vents shows that living things can adapt to any available source of energy. Also, it is now understood that today's oxygen-rich atmosphere on Earth is a relatively recent phenomenon. The first living things on Earth survived in the absence of free oxygen. The appearance of the first plants, which release oxygen as a by-product of photosynthesis, probably created an environmental crisis for other organisms.

The large gas giants of the solar system—Jupiter, Saturn, Uranus and Neptune—have usually been ruled out of consideration as candidate worlds for life elsewhere in the solar system. One reason for this is that they have reducing atmospheres, instead of oxidising. Another is that there are relatively small amounts of water. However, there may be chemical alternatives. Ammonia, for instance, has chemical similarities to water.

What are the fundamental requirements for life to thrive? We need to look for several criteria. Firstly, there has to be an adequate energy source. For the Earth, this source is solar radiation. Secondly, there has to be an environment that lies within an acceptable temperature range. Although the Earth has extremes, most of the world most of the time lies within the range of 0–30 degrees Celsius. At very low temperatures, chemical reactions slow down and even stop. At high temperatures, proteins become denatured; that is, their delicate three-dimensional structure is distorted as chemical changes occur along the protein molecule. This is a problem on Mars, where even during the day temperatures can remain far below zero. Many worlds have zones that lie within a suitable range, at least for part of the time.

Thirdly, there needs to be an abundance of essential chemical elements from which to build bodies. It is almost certainly no accident that life on Earth is based on carbon. The chemical properties of carbon make it ideal for building complex molecules of the kind needed to evolve living things.

Finally, and most problematic, there needs to be a medium rich enough in interactions and variety to allow chemical and biological evolution to proceed. One problem with the idea of organisms in the Martian soil is that the medium provides very low mobility. In contrast, the Earth has an abundance of mechanisms. There is vigorous debate over the medium in which life may first have appeared on Earth. One view is that the oceans were originally a soup of organic compounds within which chemical interactions eventually gave rise to self-replicating forms. Other scientists claim that the oceans could not have achieved a sufficient concentration of the necessary chemicals for chemical evolution to occur. Shallow water pools in which organic compounds had become concentrated would have been far richer in chemical interactions.

Perhaps the best candidate for a home for life in the solar system is the planet Jupiter. It satisfies all the criteria listed above. Solar radiation is much weaker on Jupiter— even than Mars. However, Jupiter is so massive that it is really a failed star. Its internal heat is so great that it actually radiates about four times as much energy as it receives. Temperatures at the top of the atmosphere are well below freezing, but temperatures at the centre of the planet are very high. Somewhere between these two extremes is a zone where the temperature is identical to that of the surface of the Earth. However, there is a complication. The composition of Jupiter is so different from the Earth's that it's not clear whether Jupiter has a surface at all. If it does, are there oceans of any kind? And do temperatures in the right range occur where organic chemicals can concentrate and interact?

On the other hand, evidence suggests that Jupiter has the right 'laboratory conditions' for creating the ingredients of life. Lightning has been observed in the Jovian atmosphere. Remember that the experiments by Harold Urey and others were based essentially on conditions found on Jupiter. So we can be confident that on Jupiter, where the experiment has been in progress for billions of years, there must be considerable quantities of amino acids and other building blocks of life.

To understand the origins of life we have to go back to the chemical processes that preceded the appearance of living things. How did life-like properties emerge during this period of prebiotic evolution? Perhaps the most basic characteristic is self-replication. It is generally assumed that the precursors to living things were molecules that were capable of duplicating themselves. Another relevant property is the ability to accumulate information by growing longer and longer.

Given the universal nature of genetic information in modern organisms, it is generally accepted that the chemistry of the genetic code must have been an important development during prebiotic times. The obvious candidate is DNA. But the mechanics of DNA replication are relatively complex. It looks increasingly as though DNA emerged later as an intermediate step in the replication of various forms of RNA. That is, it looks as though living organisms emerged in an RNA world as a device to assist in the chemical process of replicating RNA.

We are still a long way from understanding in detail the progression of prebiotic evolution. However, along with self-replication was the need for a mechanism that could accumulate information by accommodating longer and longer molecules. Here there is a fundamental difficulty. The problem is one of complexity. No process of information transfer is ever perfect. There is always a certain error rate. Now the problem, as Eigen and Schuster pointed out, is that self-replication is limited by an 'error threshold'. That is, for any given error rate there is an upper limit on the size of molecule. If the molecule grows beyond the critical size, then there is a cascade of errors during replication that prevent the molecule from copying itself correctly. In other words, it self-destructs.

However, Eigen and Schuster suggested that a solution to this problem existed in the form of the *hypercycle*. A hypercycle is a cyclic process in which a series of molecular species each act as catalysts for the next species in the sequence. Catalysts are chemicals that help a chemical reaction to proceed. So, in a hypercycle, chemical species A might help species B to form, species B might help species C to form, and C in turn might help A to form. The point is that each chemical species in this process might be relatively simple, so that error rates are not a problem. However, by adding extra steps in the cycle, the entire information content of the process can grow ever more complex.

It didn't take long before a number of researchers pointed out a flaw in the hypercycle theory. This is that it can easily be killed by 'parasitic' species. Suppose that at some point in the cycle one of the species catalyses not only the next member of the cycle but also another (parasitic) species. Let's call this parasitic species D. Then on each repeat of the hypercycle, the population of species D will increase at the expense of the entire cycle. Thus D acts as a dead end that eventually traps all the material involved.

A solution to this dilemma was discovered by the Dutch team of Boerlijst and Hogeweg. They showed that the spatial distribution of the chemicals is crucial. Instead of the entire 'world' cycling at once, the hypercycles became centred on various foci and produced outward spirals. Parasitic species tended to get pushed towards the outside edges of the spirals where they were removed by other processes.

What *is* life?

Scientists and philosophers have struggled with this question for generations. Among the many criteria that have been identified are:

- self-reproduction
- ability to interact with the environment
- ability to interact in non-trivial ways with other organisms

There is still a large body of opinion arguing that life on Earth is unique in the universe. To find another example, even microbes on Mars would establish whether we really are unique. The more science tackles the problem the clearer it becomes that life itself is a natural result of processes that are widespread in the universe.

People tend to assume that the forms of life we see around us are all that is possible. Without other examples to study it is difficult to know just how great the possible variations might be. In earlier chapters we saw that where direct experiments are impossible scientists are turning increasingly to simulation as a way forward.

Simulation is one of the key research methods of the new field of science called artificial life (see Chapter 3). The basic aim of alife research is to answer the question posed in the heading of this section. That is, its goal is to understand life and living processes in general. As the name implies, one test of our understanding would be to create artificial life forms. Many of the topics in this book really fall into this field of alife, which is also helping us to understand the origins of life here on earth.

Silicon Life Forms

Does life need to be carbon-based? As we saw earlier, carbon forms an astonishingly rich variety of complex molecules. Of these there is at least one, DNA, which can produce a detailed genetic code. It is not obvious that other elements can provide such a molecular code, which confines life to essentially the temperature zone of tepid water.

However, the fledgling discipline of artificial life suggests that new life forms may be created in computer environments. With IBM's Deep Blue currently the equivalent of world chess champion, the power of computers rivals that of the human brain in some domains.

Thomas Ray, in the Tierra computer program, introduced digital life forms to computer memory. These are little more than fragments of computer memory with properties of breeding, mutation and so on. While Tierra is at a very primitive stage of evolution, the drive to use evolutionary and learning systems in robotics will surely see more and more complex 'silicon' life forms.

The thing that is missing compared with real life is reproduction. We can imagine building machines that can take over the process of building machines— and perhaps ultimately leave us out of the picture. But we are still a long way from DNA, with its hundreds of megabits of genetic information in a *single molecule*.

What is consciousness?

At school we are taught about various definitions and characteristics of life. They usually include reproduction, something that our robots are a long way from doing. But most of us can probably think of other properties as being important in defining higher life forms. Of these, consciousness is the most intractable and puzzling.

Is consciousness a property of brains amenable to biological/physical explanation or is it something else, not

197

accessible to scientific enquiry? At the moment we can't say, and no accurate philosophical discussion is ever short. Let's look at some specific issues and see how they impinge on the likelihood of artificial consciousness.

An interesting approach is to imagine being able to build a mechanical nerve cell: it has the same input/output properties as a biological cell, but it's powered by a battery and uses connecting wires. We know from the work of many neuroscientists that we can insert wires or capillary tubes into cells and artificially activate them. Thus there does not seem to be any real difficulty in doing the swap. Now we keep adding artificial neurons, one by one, until our brain is entirely synthetic. Has consciousness disappeared? Extensive debate has not produced a clear answer, but readers might like to make up their own mind.

Even if we reject the notion that artificial implants will have no effect on consciousness, that there is some flashpoint beyond which consciousness vanishes, we still have to assume a very close link between the activity of the brain and consciousness: brain damage, fever, drugs, so many things that perturb brain chemistry perturb consciousness too. Even if we follow Ray Jackendoff and assume that there are separate mind–brain and mind–mind problems, we can't ignore the very tight coupling of conscious and neural behaviour.

It might be tempting to assume that some level of connectivity or size is the point at which a brain becomes conscious. Yet, as Jackendoff describes very lucidly, there is a lot of very complicated processing going on in a brain of which we are *not* conscious and never can become conscious—all the pre-processing that goes on in the sensory systems, for example. An interesting corollary here is sleep. The memories of dreams (admittedly faint and erratic) feel like conscious experiences, even though they were not.

Francis Crick, Nobel Laureate along with James Watson for discovery of the structure of DNA, has actually proposed that consciousness resides in a particular brain structure, also implicated in sleep control. But if it resides in a particular brain region, and isn't instead some global property, then perhaps that area could go wrong, get injured, leaving us a substantially functioning person but someone who is functioning more like an automaton than a person. It's hard to imagine what this might be like or even if it is possible.

Let's turn from the intractable problem of what consciousness *is* to what it is for. Again, we're no further down the track. There is evidence for it appearing in stages as we go up the animal kingdom, though. Birds will attack their reflection in a mirror or a pane of glass with no notion that this might not be another bird. A cat is more sanguine, quickly realising that its reflection is *not* another cat. We accept this since a cat will go into attack mode against a cat on the other side of a window where smell and, to a significant extent, hearing are no longer cues. But a cat is not self-aware. It cannot make deductions about itself from its reflection. Chimpanzees, on the other hand, *are* self-aware: they realise from the reflection in a mirror that a mark put on their head while they are under anaesthetic is theirs. For our part, human language takes us a further, massive, step in understanding ourselves and our relationship with the world.

With the enigma of consciousness, we come to the end of our book. It has been a fascinating journey through a landscape in which we can see the broad outlines, even grasp the richness of the scenery. But, like our first trip by car through some exciting place, we yearn to go back on foot and see it close up. So it is with complexity. It will occupy the world's finest minds for many years to come.

NOTES

1. Tom Wolfe describes this incident in detail in his book *The Right Stuff*.
2. The Analytic Engine was the first (unsuccessful) attempt to build a programmable computer. It foundered largely because early nineteenth-century technology was not capable of the precision machining required. Ada Lovelace was Lord Byron's niece. Her achievements are commemorated in the programming language Ada, which is widely used in the United States and elsewhere for military grade software.
3. As an aside, temperature is most meaningful for gases in thermal equilibrium where we get a smooth variation in the energy of individual molecules with just one peak in the number of molecules at each energy. Lasers are far from equilibrium, with a very asymmetric distribution of energies.
4. There is just a shade of doubt here. As we discuss later in the book, computation is essentially reversible, consuming no energy. Without an energy constraint, we don't know how fast we can ultimately make a computer. But for computation to be reversible we have to save all the intermediate steps. Since we physically need matter of some kind, even if we get storage down to the atomic level with so-called nano-technology, we are still defeated by the exponential growth in computing demands. There is speculation that quantum computation may allow easier storage of intervening steps, but this is far from established at the time of writing.
5. We have glossed over some of the mathematical details here. The number of steps or the time on any computer will depend on the particular method or algorithm used to solve the problem. Here we are using the simplest 'brute-force' algorithms to illustrate the point.

200

6. If the number of pixels is N and the number of levels each pixel can occupy is L, then the number of pictures, P, is L^N

7. Quantum mechanics is somewhat different, introducing additional constraints on the accuracy of measurement.

8. From D.H. Meadows, D.L. Meadows, J. Randers and W.W. Behrens III (1972), *The Limits to Growth*, Universe, Washington.

9. The naming of the different modules, usually referred to as *areas*, is not straightforward. Names given by the early anatomists have stuck and other names have been grafted on. Suffice it to say that it is somewhat deeper into the brain than the areas where most of the visual input from the eye arrives.

10. This idea underlies a major compression algorithm for images and sound, predictive coding. Instead of transmitting everything in a signal, transmit only the differences against what you would predict.

11. See J. Poskanzer (1991), *Xantfarms Simple Ant Farm for XII*. The Internet address is http://www.acme.com/software/-xantfarm/.

12. The Coriolis force acts on objects moving in a rotating system, causing their direction of motion to curve. It is commonly assumed to be the reason that water rotates in a particular direction as it goes down the plughole in a given hemisphere.

13. Creationists and opponents of evolution have often used the evolution of the eye as an example of the 'impossibility' of evolution. One of the most exciting developments of recent times has been an awareness of just how powerful evolutionary techniques really are. In fact, a paper in the proceedings of the Royal Society of London a few years ago shows how it *is* possible to evolve the lens as an eye in a quite small number of generations given some plausible assumptions.

14. Wintermute was a large and somewhat ruthless artificial intelligence in *Neuromancer*.

15. Clever Hans was a horse who could do arithmetic! He would tap a front hoof the exact number of times equal to the product of two numbers. In fact what he was doing was just tapping until he detected some cues from his trainer's demeanour that he was supposed to stop.

16. Australia's Commonwealth Scientific and Industrial Research Organisation.

BIBLIOGRAPHY

Barnsley, M. (1988), *Fractal Cities*, Academic Press.

Batty, M. and Longley, P. (1994), *Fractal Cities*, Academic Press.

Bossomaier, T.R.J. and Green, D.G. (1997), *Complex Systems*, Cambridge University Press.

Casti, J. (1994), *Complexification*, HarperCollins.

Eldredge, N. (1995), *Reinventing Darwin*, Weidenfeld & Nicolson.

Hall, N.(199), *The New Scientist Guide to Chaos*, Penguin Books.

Hodges, A. (1997), 'Turing: Alan Turing, a natural philosopher', No. 3, *The Great Philosophers*, Weidenfeld & Nicolson.

Holland, J.H. (1992), *Adaptation in Natural and Artificial Systems*, MIT Press.

Kauffman, S. (1993), *The Origins of Order*, Oxford University Press.

Kelso, J.A.S. (1995), *Dynamic Patterns*, MIT Press.

Levy, S. (1992), *Artificial Life*, Penguin.

Mandelbrot, B.B. (1982), *The Fractal Geometry of Nature*, W.H. Freeman.

Maynard Smith, J. and Szathmary, E. (1995), *The Major Transitions in Evolution*, W.H. Freeman.

Mossel, B. (1973), *Where No Road Goes*, Enterprise Publications.

Penrose, R. (1995), *Shadows of the Mind*, Vintage.

Peitgen, H-O. and Saupe, D. (1988), *The Science of Fractal Images*, Springer Verlag.

Pinker, S. (1994), *The Language Instinct*, Allen Lane.

INDEX

203